Mathematical Literacy

Why do so many learners, even those who are successful, feel that they are out-siders in the world of mathematics? Taking the central importance of language in the development of mathematical understanding as its starting point, *Mathematical Literacy* explores students' experiences of doing mathematics from primary school to university—what they think mathematics is, how it is presented to them, and what they feel about it. Building on a range of theory which focuses on community, knowledge, and identity, the author examines two particular issues: the relationship between language, learning, and mathematical knowledge and the relationship between identity, equity, and processes of exclusion/inclusion.

In this comprehensive and accessible book, the author extends our under-standing of the process of gaining mathematical fluency and provides tools for an exploration of mathematics learning across different groups in different social contexts. *Mathematical Literacy*'s analysis of how learners develop particular relationships with the subject, and what we might do to promote equity through the development of positive relationships, is of interest across all sectors of education—to researchers, teacher educators, and university educators.

Yvette Solomon is Reader in the Department of Educational Research, Lancaster University, England.

STUDIES IN MATHEMATICAL THINKING AND LEARNING
Alan H. Schoenfeld, Series Editor

Mathematical Literacy

Developing Identities of Inclusion

Yvette Solomon

Routledge
Taylor & Francis Group

NEW YORK AND LONDON

First published 2009
by Routledge
711 Third Avenue, New York, NY 10017, USA

Simultaneously published in the UK
by Routledge
2 Park Square, Milton Park, Abingdon, Oxon OX14 4RN

Routledge is an imprint of the Taylor & Francis Group, an informa business

© 2009 Routledge, Taylor and Francis

Typeset in Minion by
Swales & Willis Ltd, Exeter, Devon

Library of Congress Cataloging in Publication Data
Solomon, Yvette.
Mathematical literacy: developing identities of inclusion/Yvette Solomon.
        p. cm.
    Includes bibliographical references and index.
    1. Mathematics—Study and teaching.   2. Mathematical readiness.   I. Title.
    QA11.2.S64 2008
    510.71—dc22   2008016897

ISBN 10: 0–8058–4686–7 (hbk)
ISBN 10: 0–8058–4687–5 (pbk)
ISBN 10: 0–2038–8927–4 (ebk)

ISBN 13: 978–0-8058–4686–7 (hbk)
ISBN 13: 978–0-8058–4687–4 (pbk)
ISBN 13: 978–0-2038–8927–5 (ebk)

Contents

Preface

In this book I have listened to what mathematics learners say about their experiences of doing mathematics in classrooms from primary school to university: what they think mathematics is, how it is learned, what their lessons are like and what they feel about it. Inspired by Jay Lemke's question "How do moments add up to lives?", I have tried to make sense of what these students say in terms of how our individual trajectories through mathematical learning are the interwoven products of the many moments which add up to our mathematical lives or identities. In particular, I have aimed to understand why so many students, even successful ones, have negative relationships with mathematics. To do this, I have explored what they say about learning mathematics from a number of complementary angles: in terms of their engagement in a community of practice; in terms of their epistemologies of mathematics; and in terms of the discourses of gender, pedagogy and mathematics itself which frame their mathematical lives and make some identities more available than others.

In developing an account of why so many students feel excluded from mathematics, and conversely what is needed in order to develop identities of inclusion, I have focused on the issue of access to mathematical discourse itself and to the way in which it makes meaning. Hence this book is called *Mathematical Literacy*, since meaning-making through language is a necessary part of participating in the practice of mathematics, and it is this meaning-making which appears to be so out of reach to so many people for a number of reasons.

The book builds on a range of socio-cultural theory and research which focuses on community, knowledge and identity, and on a large body of research in the area of mathematics learning and education which has been undertaken in the UK and the USA. Two particular areas stand out in the way I have explored these issues: one concerns the relationship between language, learning and mathematical knowledge, the other that between identity, equity and processes of exclusion/inclusion. I draw on these throughout the book as I work through the issues which are raised by the students who are its focus— somewhat unusually, this book spans a wide age range, beginning with children who are 9- or 10-year-olds at primary school, moving on to secondary school students aged 11–12, 13–14 and 14–15, and finally on to undergraduates in the British university system. This has enabled me to draw parallels between the students, and to speculate on "how moments become lives."

This book is intended for researchers and teacher educators, and university educators. Although it is about mathematics learning, and will be of particular interest to those working in this field, it will also be of interest to researchers in the area of applications of theories of communities of practice, learner identities, and academic literacies. While this book focuses on mathematics literacy, its analysis of how learners develop particular relationships with the subject, and what we might do to promote equity through the development of positive relationships, is applicable more generally, across all sectors of education.

Part I: Inclusion and Exclusion

In Part I, I tell the stories of a wide range of mathematics students. I am concerned to explain primarily how their identities as learners are shaped by their experiences, and the ways in which their responses to those experiences support and are supported by particular beliefs and discourses. I show that there is evidence of repeated positionings which generate identities of inclusion and exclusion which follow noticeable patterns relating to class and gender. The identities which learners take up in these circumstances are connected to their participation in the practice of mathematics in terms of their epistemologies of mathematics and their relationships with teachers—as resource, for those who are potentially included in the practice, or as authority, for those who are marginalized. While some students develop positive relationships with mathematics, many do not, and this is visible even at undergraduate level.

I set the scene in Chapter 1, which lays out the problem of developing mathematical literacy within the context of mathematics as a social and cultural activity. I introduce issues in the language of mathematics, in discourses *about* mathematics, and in the nature of classroom talk. I also introduce a framework for theorizing about mathematics literacy, in terms of Wenger's communities of practice model, his focus on identity in communities of practice, and also the concept of personal epistemologies. I develop and add to these throughout the book as particular patterns emerge from the data analysis. Finally in Chapter 1, I introduce the data itself and the contexts in which it was gathered.

In Chapter 2, I begin to explore the ways in which classroom cultures develop, and I review the literature on pedagogy in mathematics classrooms, particularly that concerned with the practice of "ability grouping" or "setting." This is very important, since, while setting is strongly criticized in the educational research arena, it is an almost unquestioned practice in England and Wales on a school, community and political level. It has a long-established history in secondary school mathematics teaching, and has in the last decade or so become established practice in some form or other in primary schools. As I will show in later chapters, the impact of setting on identity is considerable. I also consider the role of class, ethnicity and especially gender in the development of classroom interaction patterns, illustrating these with data drawn from primary years classrooms.

In Chapter 3, I move on to look more closely at mathematics identities in the secondary school years. In what is a fairly typical pattern, the students at Northdown School are taught in heterogeneous groups in Year 7, their first year at the school, moving to a setting system in subsequent years. With the Year 7 students as a baseline, I explore the impact of setting in Years 9 and 10; I conclude that it makes a difference to the ways in which the students experience and respond to their mathematics lessons, and to their perception of their relationships with teachers. An important aspect of the students' response to setting and to learning mathematics is the way in which they are positioned by discourses of ability. In addition, gender emerges as a major complicating factor in the experience of mathematics classes and the development of students' relationships with the subject.

In Chapters 4, 5 and 6, I turn to look at undergraduate experiences of mathematics teaching. As a group of students they are "good at mathematics" but for many their perceptions of themselves as learners and as mathematicians are startlingly negative. In Chapter 4, I review the literature on students' experiences of learning mathematics at university level, and explore the nature of undergraduate identities in terms of modes of belonging in the mathematics community of practice: I show that, in a complication for Wenger's theory, a gender difference emerges in how students relate to mathematics such that women apparently have more engaged identities and yet are more anxious and self-critical about their mathematics abilities. This is a theme which re-emerges in Chapter 5, where I consider the beliefs about mathematics held by the same group of students. I show that their perceptions of learning and teaching relationships inscribe particular identities of participant and non-participant in the central practices of mathematics, as a case study of proof shows. What is striking about the undergraduates is their relationship to authority, which seems to be very similar to the secondary school students, and their continued self-positioning with respect to discourses of ability and an associated emphasis on speed. Emerging from the analysis is a concept of "fragile identities," exemplified particularly in the women's narratives of engagement in mathematics coupled with a vulnerability in terms of the pervasiveness of these discourses.

In Chapter 6, I explore the concept of fragile identities further with a new group of students, second- and third-years in two different universities. I explore their "figured world" and the identity positions within it—I show how they position each other and themselves in terms of "genius" or "geeks," and also how they position their tutors as particular types of people. What is interesting about this group is the way in which they have developed their own collaborative culture of being "good at mathematics," using mathematics support centers in their two universities, and so "creating spaces" to be. On an individual level, some can be seen to be creating spaces in the sense of refigured selves as they make decisions about how to respond to their experiences of mathematics learning.

Part II: Developing Inclusion

In Part II, I explore the options for addressing the problem of exclusion from mathematics. I focus on the question of what is it about mathematics that creates exclusion, reviewing research and theory concerning class, culture and gender differences in mathematics and their proposed solutions. Building on a primarily linguistic analysis, I suggest that identities of inclusion can only be developed within an inclusive pedagogy which makes the discourse of mathematics—and therefore its central practices—visible, and enables teacher–learner relationships which foster opportunities for discussion and challenge.

In Chapter 7, I review research on exclusion from mathematics, beginning with the ways in which issues of class and culture are played out in mathematics so that, for instance, linguistic and cultural capital play a role in how learners are positioned with respect to access to the mathematics register in general, and the kinds of textbooks that they are given in particular. I then turn to a consideration of gender and the argument, based in difference feminism, that the pedagogy and content of traditionally taught mathematics does not suit girls' and women's ways of knowing and relating to subject knowledge. Looking back through the data which forms the focus of Part I, I argue that, while it has a surface emphasis on right answers, formality and certainty which is reflected in the discourses of natural ability and speed which circulate within the classroom community, school mathematics also supports a creativity and connection-making which some students access and incorporate into a positive mathematics identity.

I explore these ideas further in Chapter 8, drawing on parallels in literacy education to develop an account of what is required to support an inclusive mathematics literacy. I argue that the language of mathematics is not arbitrary but is central to its practices of meaning-making, and therefore that an inclusive mathematics pedagogy needs to make this language explicit. I examine the implications of this for teacher–learner relationships, arguing that the complexity of mathematical language and the teacher's necessary role as epistemic authority requires a relationship which is based in possibilities for argument and challenge on both sides.

Finally, I explore what all this means for practice in Chapter 9. Drawing on a range of literature, primarily from equity initiatives in the USA, I focus on what is involved in developing inclusive and positive relationships with mathematics. I outline the particular challenges that this brings, in terms of the role of the teacher-expert and the dilemma of when to intervene versus allowing learners to develop their own arguments, and in terms of student identities and self-positionings which may in fact involve resistance to nontraditional teaching. I explore the implications of the fact that developing classrooms which theoretically enable better access to mathematics is only one side of the story—changes also have to coincide with individual students' sense of who they are.

Acknowledgements

This book is the product of many years of researching, thinking about and actually doing mathematics. It draws on ideas that I originally developed in the late 1980s, when I was first concerned to understand what it might mean to say that mathematics is a social practice. Since then, I have continued to think through the implications of this stance, helped along the way by a large number of people whose work, conversations and comments have contributed to my thinking in various ways; I would like to thank as many of them as I can here. First, of course, are the students who generously gave up their time to talk to me about doing mathematics, sharing their anxieties and their successes. I also owe a very special debt to Laura Black, not only for allowing me access to her data, but also for the inspiration that I have drawn from many years of working with her. Thanks also go to Heather Mendick, for her tireless pursuit of different ways of looking at things, and everything that I have learned as a result. Both have read and commented on various parts of the book and helped me to improve it. Together with Margaret Brown and Melissa Rodd, our co-organizers of the Economic and Social Research Council-funded seminar series *Mathematical Relationships: Identities and Participation* which ran in 2006–07, we have debated and learned a huge amount and I owe a lot of the thinking in this book to this experience. I would also like to thank the many participants in the series, but particularly Tamara Bibby, Jo Boaler, Tansy Hardy, Steve Lerman, Candia Morgan, Hilary Povey, Barbara Jaworski, Jim Ridgway and Jenny Shaw for their contributions and later collaborations and conversations.

A number of other people and groups have played a part in making the book possible. I am particularly grateful to Tony Croft and Duncan Lawson at SIGMA, the Centre for Excellence in Mathematics and Statistics Support at Coventry and Loughborough Universities, for inviting me to participate in research there, and to Peter Samuels for his interest in my work. I have also gained immensely from my involvement with the Socio-Cultural Theory Interest Group at Manchester University, and discussions there with Laura Black, Pauline Davis, Valerie Farnsworth and Julian Williams among others. The two conferences organized by this group in 2005 and 2007 provided opportunities to work through the ideas that are expressed in Chapters 4 and 6 of this book, and Barbara Crossouard's input on both occasions was much appreciated.

I am enormously indebted to the series editor, Alan Schoenfeld, and to James Greeno, Melissa Gresalfi and Victoria Hand who all wrote superb responses to the first draft of this book; between them they provided me with challenges and new insights which strengthened what I wanted to say. I would also like to thank Jo Boaler and Anna Sfard for their positive and helpful comments on the initial proposal and for our more recent conversations. I am very grateful to Naomi Silverman, my first editor at Lawrence Erlbaum Associates, for her patience and encouragement as I struggled to find the time to write, and to Catherine Bernard who took over towards the end. Thanks too to my colleagues at Lancaster University for their support, particularly Joanne Dickinson, Anne-Marie Houghton, Carolyn Jackson, Colin Rogers and Jo Warin.

Almost exactly in parallel with writing this book, I studied for an Open University degree in mathematics and statistics. Why I decided to do this is rather a long story, but I learned a lot, and not just about mathematics. It made me feel more confident in writing this book, and enabled me to experience some different relationships and indeed different feelings. Special thanks go to my OU tutors, particularly Peter Gregory, Hilary Short and David Towers.

Lastly, I couldn't have done this without the support of my family. Not only have they tolerated my neglect of them as I worked on this book, my job at Lancaster and my OU degree, inflicting this latter in particular on them even on summer holidays, plus my angst over just about everything, but they have never ceased to encourage me to do it all. So thanks to John O'Neill for all the conversations over the years, to Bridie O'Neill for hanging on in there, and to Rosie O'Neill for all that cooking. This book is for you.

Parts of this book have previously appeared elsewhere, and earlier versions of Chapters 3, 4, 5 and 8 were published as follows.

Chapter 3: Solomon, Y. (2007) Experiencing mathematics classes: Ability grouping, gender and the selective development of participative identities, *International Journal of Educational Research* 46 (1/2) 8–19, copyright (2007), Elsevier.

Chapter 4: Solomon, Y. (2007) Not belonging? What makes a functional learner identity in the undergraduate mathematics community of practice?, *Studies in Higher Education* 32:1 79–96, Taylor & Francis, http://www.tandf. co.uk/journals/

Chapter 5: Solomon, Y. (2006) Deficit or difference? The role of students' epistemologies of mathematics in their interactions with proof, *Educational Studies in Mathematics* 61:3 373–393, Springer Science and Business Media.

Chapter 8: Solomon, Y. (1998) Teaching mathematics: Ritual, principle and practice, *Journal of Philosophy of Education* 32:3 377–390, Blackwell Publishing.

Solomon, Y. & O'Neill, J. (1998) Mathematics and narrative, *Language and Education* 12:3 210–221, Multilingual Matters.

The original publishers have kindly given permission for this material to be used again here.

Part I
Inclusion and Exclusion

1

Formulating the Problem
Identifying Mathematical Literacy

Attempts to understand success—and more often failure—in mathematics education have drawn on a number of different approaches in the fields of psychology, sociology and linguistics. The range is vast: it includes cognitive modeling of arithmetic algorithms; cognitive constructivist theories of concept development; social-psychological analyses of motivation patterns; psychoanalytic explanations of emotional responses to mathematics such as fear and anxiety; feminist and discourse-based analyses of exclusion and inclusion; and social constructivist approaches which focus on cultural mediation and apprenticeship. Here, I take the position that mathematics is a social practice, a standpoint that I developed in an earlier book (Solomon, 1989). My argument then was that Piagetian cognitive constructivist conceptions of mathematics and learning could not explain how mathematical ideas could develop in an individual learner alone, and that mathematical knowledge had to be seen as intrinsically social in order to make sense of the learning we do. My aim now is to explore this latter issue further by looking more closely at the ways in which learners develop identities of participation or marginalization within formal learning contexts, and how they consequently gain—or fail to gain—access to the central meaning-making practices of mathematics—that is, how they become mathematically literate. In this chapter, I set the scene for this exploration by reviewing theory and research which describe mathematics and mathematics learning as inherently social and cultural activities, and introducing the initial theoretical framework which underpins my analysis of what it means to become mathematically literate.

Mathematics as a Social and Cultural Activity

Theory and practice which recognizes mathematics as a social and cultural activity draws on three major strands of research involving the role of language. The first of these focuses on the language of mathematics itself, ranging from observations about its basic vocabulary to the ways in which it makes meaning in its grammatical forms. Specific issues are uncovered by this research, such as the difficulties for learners presented by the complex relationship between everyday and mathematical language and the role of metaphor in mathematical representations and modes of argument. The second strand is closely connected to the first: it concerns the apprehension of discipline-based ground rules as carried within the written and spoken mathematics discourse, and the extent to which learners are able to recognize and control the discourse of mathematics with respect to its relationship with everyday practices, for example. It also concerns the role of discourses *about* mathematics: these discourses are key to beliefs that individuals hold about mathematics and how it is learned and by whom. The third strand draws on Bernstein's identification of the sociological context of classroom cultures, and on the neo-Vygotskian tradition in terms of a concern with the role of language as a tool in mathematics learning. This research points to issues in classroom cultures and the ways in which interactions between teachers and pupils shape access to mathematical understandings. Clearly, these three strands are intertwined, and I move between them in this book.

I will begin this section by recounting an episode from my observations in a British state primary school classroom of 10-year-olds—they are in Year 5, their penultimate year in this school. It illustrates many aspects of the social and cultural nature of mathematics learning and I will use it to identify a number of issues that I address in this book, and to outline its theoretical framework.

The class is arranged, as is typical in British primary school classrooms, in tables shared by children with similar attainment levels, and they are just about to embark on their daily mathematics lesson. The teacher, who I will call Mr. Wilson (all the names of people and places in this book are pseudonyms), introduces the topic. He tells me that he has picked a set of word problems (and for later lessons, a sequence problem and geometric problems) as a way of fostering a "feel for mathematics." He also has a more specific agenda, which is to encourage the children to gain marks for method in their mathematics Standard Assessment Tests (SATs). Although they still have a year to go before they take their Key Stage 2 statutory SATs, they will take an internal SATs test in the current year, and one of the props for this lesson is a Level 6 "extension paper"; this is a paper given at the time of this observation to higher attainers identified by the school, taken in addition to the standard paper, which only assesses Levels 3 to 5. To the class he says:

> We are using words in arithmetic and recognizing different *processes*: note the difference between a question which asks you to . . . cut up 39 cm of ribbon into 6 equal pieces, share 39 marbles between 6 people, or work out how many egg boxes you need for 39 eggs if each box holds 6 eggs. We use the same arithmetic process but there are three different answers.

He rehearses with the pupils the different answers to each question type and how these are justified. Then he enters into a question and answer routine with the whole class:

Mr. Wilson: Multiply six and another number to get fifty-four.
Pupil: Nine.
Mr. Wilson: Why nine? What was the *process*?
Pupil: Divide.
Mr. Wilson: Yes, divide. Look at the Level 6 SATs paper marks boxes. *(He holds up a Level 6 SATs paper.)* It says "show your working, you may get a mark." SHOW YOUR WORKING!

Having introduced the topic in this way, with considerable emphasis on "showing working" in order to gain test marks, Mr. Wilson sets the children to work on some word problems. He presents the class with a typed sheet of some 20 problems, and they get on with these alone or in pairs. An early question on the sheet asks: *If a family of 4, eating three meals a day, eats 12 meals a day between them, how many meals will the family eat in a leap year? A leap year has 366 days.* Julie, who is one of the lower attainers in the class, turns to me for help.

Julie: The question is wrong—a leap year has 360 days. . . . It's twelve divided by four.
YS: How do you know?
Julie: Because the last two [questions] were "divides."
YS: Well, it doesn't have to be divides does it? Read the question again carefully, what does it ask you to work out?
Julie: *(reads out question, slowly)* How many! It says "how many," that means . . . *(looks up at posters on wall featuring the symbols for division, addition, multiplication and subtraction and their associated words—"share," "difference," "sum," etc)* . . . Look that means "lots of," "many" means "lots" so it's a "times". . .
YS: So what are you going to multiply to answer the problem?
Julie: Twelve times four.
YS: How do you know that?
Julie: Well, the middle number is the one you multiply.

Paul, on the other hand, speeds through the problems, getting to the following question very quickly: *Pencils come in packets of 6. There are 36 children in Lisa's class. If the teacher buys 6 packets, will she have enough for one pencil for each of them?* I ask him some questions about his answer.

Paul: *(writes "36 × 6 = 216" in his exercise book.)*
YS: Can you tell me how you decided it was 36 times 6?
Paul: It just is.
YS: Well, have another look at the question and see if you can talk it through . . . *(he changes his written answer to 36/6)*
YS: Think about what Mr. Wilson said, show your working, can you do that? What's the first thing you have to work out?

Paul: *(no response)*
YS: Well, how many pencils does the teacher need?
Paul: Six times six. She needs thirty-six.
YS: Yes, so what do you have to do to show that she has enough pencils, to answer the question?
Paul: *(no response—looks puzzled)*
YS: You could write down, couldn't you, that she has 36 pencils and so she has enough for the 36 children in the class, to show your working?
Paul: *(changes written answer to 6 × 6 = 36)*

Julie and Paul provide contrasting and familiar pictures of mathematics learners. Julie presents the well-known phenomenon of the learner who cannot identify irrelevant information and who struggles with the interpretation of mathematical symbols and sentences; she quickly asks for help. Paul, on the other hand, epitomizes the student who wants to move quickly through the problem sheet and get the answers right; he is not very interested in talking about what he is doing. Clearly, Julie is not as competent in answering the questions as Paul, but, as Sfard (2006, p. 156) notes, "knowing what children usually *do not* do is not enough to account for what they *actually do.*" Rather than seeing Julie's approach to mathematics in terms of absences in her acquisition of knowledge, a more "participationist" view (see Sfard, 2006, p. 153) seeks to understand how she engages in mathematics as a social practice in terms of the various cultural resources and discursive positionings that she brings to the situation. Sfard articulates this strong participationist view thus:

> According to this vision, learning to speak, to solve mathematical problem[s] or to cook means a gradual transition from being able to take a part in collective implementations of a given type of task to becoming capable of implementing such tasks in their entirety and on one's own accord. Eventually, a person can perform on her own and in her unique way entire sequences of steps which, so far, she would only execute with others. (Sfard, 2006, p. 157)

Identifying with mathematics in this view is a question of movement from collective action towards a position of individual action, but still with regard to its discourse rules—what Sfard calls its "endorsed narratives" (p. 163)—since participation in mathematical communication is central to its practice. On this reading, then, we may see Julie and Paul not as disconnected individuals, one "good at mathematics," one not, but as actors within the same community of practice, taking up differing positions within it.

The next step, then, is to explore the nature of that practice in terms of the enactment of mathematics within the classroom community that Julie and Paul find themselves in. Returning to my observations in the classroom itself, it is perhaps surprising that, given Mr. Wilson's emphasis on the upcoming tests, the instruction to "show your working" did not change the children's problem-solving behavior. However, if, as Sfard (p. 162) suggests, "Learning mathematics may now be defined as individualizing mathematical discourse, that is, as the process of becoming able to have mathematical communication not only with others, but also with oneself," we may see Paul's "it just is" as part of the same phenomenon as Julie's search for clues: in their different ways, they struggle to take control of the discourse. We can also see other discourses at play in this classroom: part of their resistance to the idea of breaking down the problems either verbally or in writing was that the children expected to either "see" the answer or not; consequently some of them simply marked out certain questions as "too hard for them." Paul's emphasis on speed, and his impatience with the pencils problem, appears to be part of this equation of being able to do mathematics with finding answers quickly; he is one of "the bright boys" identified by Mr. Wilson when he talks to me after the lesson, in danger of disaffection if they are allowed to get bored—his solution is to start teaching mathematics in "ability groups."

The institutional context and discursive positionings of the mathematics classroom come to the fore here, bridging the issues of identification and practice. My major concern in what follows will be to explore the ways in which individuals come to develop identities of participation and non-participation, and how they draw on cultural models and resources to develop their own particular narratives of doing mathematics. In this sense I am concerned to understand the interplay of identity and agency in the "figured world" of the classroom (Holland, Lachiotte Jr., Skinner, & Cain, 1998). So, for example, Mr. Wilson's reference to the "bright boys" invokes familiar discourses of gender and ability which may operate as powerful constraints on what positions are available to learners, as Gee (2001) suggests. The possibility of gendered and classed "designated identities" as mathematics learners (Sfard & Prusak, 2005) and children's reaction to "hard" problems are similarly aspects of their self-positioning within the available discourses, which include powerful discourses *about* mathematics. There is also an emotional aspect to these positionings of self: identifying as "good at mathematics" involves particular investments which take place within the interpersonal context of learning, and the role of teacher–pupil relationships is a crucial aspect of this socio-emotional back-ground. In terms of pedagogic practice itself, it is possible to see how Mr. Wilson's own perception of the nature of mathematics influences his teaching; despite his advocacy of teaching a "feel for mathematics," he casts the children's difficulties in terms of a lack of learned facts, to be addressed by "going back to basics." Against the explicit context of the audit culture and its impact on teaching and assessment (see for example Morgan, Tsatsaroni, & Lerman, 2002), his desire to pass on his own liking for mathematics, and his identity as a good mathematics teacher are compromised.

My aim in this book is to show how these multiple issues contribute to the development of learners' relationships with mathematics and their access to it as a social practice. I begin in this chapter with an examination of the discursive nature of mathematics and the links between identity and practice.

The Language of Mathematics: Making Meaning Through Discourse

The language of mathematics is complex and dense, and its relationship to everyday language is far from straightforward. When embedded within the further complexity of pedagogic discourse structures, mathematical language begins to present a major challenge for learners and indeed for teachers. Research in this area has developed considerably from initial observations about the potential for confusion due to specialist vocabulary uses, notation and symbolism, and the role of metaphor and modeling, to recent more sustained linguistic and semiotic treatments. Thus studies of language in mathematics education have shown the significance of the fact that numbers, arithmetic and mathematics use a particular vocabulary system and syntax (see for example Durkin & Shire, 1991); that mathematics has its own register (Halliday, 1978; Pimm, 1987); that schoolroom mathematics and mathematics texts have their own discourse (Dowling, 1998; Morgan, 1998; Walkerdine, 1988); and that mathematics can in fact be seen as a complex semiotic system (Lemke, 2003; O'Halloran, 2000, 2003, 2005, 2007).

I will not attempt an exhaustive survey of these developments in our understanding of the language of mathematics here (but see Schleppegrell, 2007, for an accessible overview), although I return to some of these themes in Part II. I will limit my discussion instead to aspects of mathematics and language that are central to my concern with the development of identities of inclusion and exclusion, making some initial points about the mathematics register in terms of relationships between the everyday and the mathematical lexicons, the role of metaphor and in particular grammatical metaphor, and—in the following sections—about the classroom production of mathematics. While these aspects of mathematics are central to the way in which it makes meaning, they can also make the task of learning mathematics difficult, although not equally so for all learners. However, I want to suggest that it is not the linguistic complexity of mathematics *per se* that creates this difficulty, but, rather, the *interaction* of these features with the classroom practices that I will describe, in Chapter 2 in particular. The narratives of doing and being good (or bad) at mathematics which I will recount in later chapters appear to have their roots in this interaction and the ways in which pedagogic practices do, or do not, enable participation in its discourse. As Sfard (2006, p. 161) puts it: "The membership in the wider community of discourse is won through participation in communicational activities of any collective that practices this discourse." In this sense, then, identifying oneself as good at mathematics relies on being part of the mathematics collective, and that means being able to "talk the talk," as

Lave and Wenger (1991) showed with members of Alcoholics Anonymous and as Holland et al. (1998) argue more generally with respect to "self-authoring." Mathematical literacy is, then, an intrinsic part of an identity of inclusion in mathematics—"meanings precede us and *we are constituted* within language and the associated practices" (Lerman, 2001, p. 87, my emphasis).

What is involved in talking the talk of mathematics extends well beyond a grasp of its technical lexis to the ability to use linguistic form to manipulate ideas. This can be understood if we take a functional linguistic[1] viewpoint such as Halliday's:

> It is the meanings, including the styles of meaning and modes of argument, that constitute a register, rather than the words and structures as such. In order to express new meanings, it may be necessary to invent new words; but there are many different ways in which a language can add new meanings, and inventing words is only one of them. We should not think of a mathematics register as consisting solely of terminology, or of the development of a register as simply a process of adding new words. (Halliday, 1978, p. 195)

One way in which the register makes meaning is through the use of conceptual metaphor—a means of concretizing the abstract, and a reflection of our cultural thinking, according to Lakoff and Johnson (1980). Pimm (1987, pp. 93ff) lists a number of examples of extra-mathematical metaphor ("an equation is a balance") in contrast to the use of "structural metaphor" in mathematics, whereby the analogy in question is drawn from within mathematics itself rather than the everyday world ("spherical triangle"). Lakoff and Nuñez (2001) show that the argument structure of proof relies on another linguistic device—metonymy—the use of an easily perceived aspect of an object or situation to stand for the thing as a whole, or a place to stand for an event as in "Watergate changed American politics" (Jay, 2003, p. 323). They argue that what they call "the Fundamental Metonymy of Algebra" (p. 74), depends on the substitution of roles for individuals (as in the generic "pizza boy," who can be any individual fulfilling the role of pizza delivery) and so enables a move from concrete case to general thinking as in "$x + 2 = 7$." Here, "x" is the notation for a role which stands for an individual number, and "it is this metonymic mechanism that makes the discipline of algebra possible, by allowing us to reason about numbers or other entities without knowing which particular entities we are talking about" (Lakoff & Nuñez, 2001, pp. 74–5). As Jay points out, however, and as Lakoff and Nuñez's example makes plain, while metonymy enables inference making, such inference will not be understood if speaker and listener do not share common knowledge and assumptions.

While metonymy supports central modes of argument, the use of grammatical metaphor in mathematics serves to create meaning by enabling congruent (i.e. everyday) linguistic representations to be reconfigured in new ways. So, for instance, instead of representing events through verbs, they are presented as nouns, and logical relations are presented as verbs; the resultant representation does not map directly on to everyday experience and so has new meaning. Veel (1999, p. 194) describes how grammatical metaphor has specific functions in mathematics: it enables the creation of quantifiable entities—it turns events or qualities into things which can then be counted. Thus while the congruent representation of change as a verb is limited ("it changes a lot"), the re-construal of change as a noun enables precise quantification ("a 50% change"). These entities can then be combined to create further complexity ("the change differential according to gender"). Grammatical metaphor also enables the reification of mathematical activities as topic areas, so that the act of multiplying becomes "multiplication," with connotations of a grasp of principle ("he understands multiplication") versus operational competence ("he can multiply").

Mathematics makes heavy use of a number of other linguistic devices which make meaning, including two types of relational clauses which classify and define mathematics, and also bridge between its linguistic and symbolic representations. The first of these makes non-reversible attributions which classify objects and events within types—(e.g. "a square is a quadrilateral")—as Veel (p. 195) suggests, "these clauses render explicit to students the organization of uncommonsense knowledge in mathematics and play an important role in apprenticing students into mathematical knowledge." The second type uses a reversible identifying relation to define technical terms in less technical language—(e.g. "the mean score is the sum of the scores divided by the number of scores")—which not only defines but also bridges between language and symbol, providing a device which "parallels the algebraic formula

$$\bar{x} = \frac{\Sigma x}{n}$$ " (p. 196). Finally, Veel points to the use of nominal groups which expand

on the meaning of an object by indicating qualities which underline its relationships to other objects, or by specifying a range of meaning (e.g. "the volume of a rectangular prism with sides 8, 10 and 12 cm") (p. 197). In this example, the object in question is a prism, volume is its quantifiable attribute, it is classified as rectangular and restricted in terms of its specified dimensions.

Overall, "a striking feature of mathematical language is the way it builds up hierarchies of technicality . . . . Each successive level of technicality takes the language user one step further away from any 'congruent' or 'everyday' construal of meaning" (Veel, 1999, p. 199). Clearly, the linguistically hierarchical nature of mathematics—and the rapidity with which it is built up in the teaching of mathematics—presents the learner with a particular challenge in talking the talk of mathematics. As I have noted above, however, I aim to show that the problem of access to mathematics lies not in this linguistic complexity as such but in its interaction with classroom cultures and practices and their joint effect on identities of participation. There are two inter-related issues here: one concerns how pedagogy attempts to bridge from everyday experience to classroom mathematics, and the other is the discourse of pedagogy itself. I shall consider each in turn.

Access to Mathematics: A Bridge from Everyday Experience to the Classroom?

This brief overview of research into the mathematics register illustrates that it is not only complex and demanding for a learner as they build on their everyday understanding of language to a mathematical understanding, but also that mathematical meaning is constructed within the language itself. Thus Pimm (1987) shows that full participation in mathematics constitutes being able to use not just individual technical terms but also phrases and modes of argument—being a mathematician involves speaking like a mathematician. This point is also made by Morgan (2005) with direct relevance to the issue of inclusion and exclusion. She notes that recent UK National Numeracy Strategy guidance for mathematics support (DfES, 2000) entitled *Mathematical Vocabulary*, makes assumptions about mathematics which underpin differential access for students deemed to be of "higher and lower ability." The guide is based on the premise that mathematical language is solely comprised of specialist words which can be simply and essentially defined, but as Morgan shows, professional mathematical uses of definition are characterized by "a relationship between definition and concept that appears dynamic and open to manipulation and decision making by mathematicians" (p. 107) and indeed this is reflected in mathematics texts aimed at higher attaining pupils but not those written for lower attainers. Her analysis, based on Halliday's system, shows that while an intermediate text on trigonometry "lays down a set of absolute and unques- tionable facts to be accepted by the student-reader," a "higher ability" text "allows uncertainty and alternatives, opening up the possibility that the student-reader herself might choose between the two definitions . . . [she] is thus constructed as a potential initiate into the practices of creative and purposeful definition that academic mathematicians engage in" (p. 114).

In overlooking the complexity of just this one aspect of mathematics, the National Numeracy Strategy guide has two important effects: firstly, it presents an inaccurate and, as Morgan says (p. 115), a restricted view of mathematics; secondly, and arguably more importantly, it obscures an important characteristic of mathematics which is in fact acted out in "higher ability" classrooms. As I will argue in later chapters, it is this tendency in mathematics teaching to hide central practices that engenders differential access. As Morgan points out:

. . . mathematical language consists of more than just specialist vocabulary. Learning to engage in mathematical discourse thus involves learning more than definitions of mathematical words. ...[the] formation [of definitions] and their incorporation into mathematical arguments are fundamental mathematical activities that take place in language. Induction into mathematical practices must involve students in developing ways of speaking and writing that enable them to engage in these activities. (p. 115)

One of the major issues which Morgan's research alerts us to is that of differential access to mathematics for different groups of learners. In attempting to analyze and explain the nature of the relationship between social class and underachievement in mathematics, a number of writers have argued that the content of mathematics is biased towards middle-class learners. So, for example, Cooper and Dunne (2000) report that working-class students perform as well as middle-class students on decontextualised tasks, but drop behind on embedded tasks despite the fact that these are commonly assumed to be "easier" as a result of their embeddedness. As Zevenbergen (2001) argues, embedding mathematics into the context of saving for a car, calculating depreciation costs and so on does not necessarily create a gain in meaning for working-class students who have little experience of such strategies. Of course, one might argue that many—or most—middle-class young people will, equally, have little experience of saving and budgeting. Given Tizard and Hughes' (1984) data on informal learning contexts, working-class children could in fact be advantaged by such an embedding, living as they do in families where financial issues are necessarily more visible.

Which way round it is does not matter though: as Cooper (2001) points out, the crucial issue is the fact that classroom mathematics problems are not supposed to be treated as real problems. Thus the explanation of under-achievement needs to be more complex than one of match and mismatch between the problem and the learner's experience: attempts to make mathematics more accessible by changing its content are missing the point and, as I will show in Chapters 7 and 8, may in fact compound the problem. The difficulty is not on the level of the cultural/linguistic content but on the level of the form of the classroom mathematics discourse. Thus a concern with "what context" might be misplaced because successful mathematics students know that the context of a problem need not resemble *anyone's* reality—in Bernstein's (1990, p. 216) terms, it is part of the "imaginary discourse" of mathematics instruction. This is demonstrated by Lerman and Zevenbergen's (2004) discussion of the socio-political context of the mathematics classroom: learners need not only to recognize the context in the sense of its power relations, but also to access the realization rules which enable them to "speak the expected legitimate text" (Bernstein, 1996, p. 32). Dealing with embedded mathematics problems therefore requires a certain level of participation in the practice of mathematics pedagogy, the ability to "spot" the problem *as* a mathematics problem, and to approach it accordingly. Beyond the classroom, mathematics has its own disciplinary aims and methods, and the apprehension and appropriate use of these is not necessarily made easier by its situation in a "realistic" context, as the examples of both Julie and Paul in my opening vignette show.

We can return to the issue of metaphor here in order to illustrate the point, this time by paying attention to the issue of *pedagogic* metaphor. Mathematics teaching makes considerable use of simple lexical metaphor—for example, early arithmetic teaching borrows from real-world experiences in its use of ordinary language such as "having" and "owing," "borrowing" and "carrying," or we can say "a function is a machine," or "an equation is a balance." Nolder (1991) notes that such metaphorical use serves a number of purposes for both teachers and learners, making it possible to talk and think about a concept, or to relate it to what is already known. However, it can of course distort perception, or can confuse because it does not apply in all situations (Nolder's example is the fact that multiplying to make numbers "bigger" breaks down in the case of fractions), or it can be over-extended and outlive its usefulness. As Lakoff's work shows, shared knowledge and beliefs are crucial in the understanding of metaphor. This is demonstrated in an example described in Walkerdine (1988), in which the teacher introduces subtraction using the metaphor of a shopping game: the children are supposed to "buy" items illustrated on cards with prices of anything up to 10 pence, and calculate the "change." They are to buy only one item at a time, starting again with a new set of money (10 "pennies") for each one. The mathematical task is to calculate the subtraction operation $10 - x = ?$ using their coins as concrete symbols which can be rearranged in piles, and then record the "change" from 10p in the form of a written subtraction sum. The teacher's intention in this very common type of activity is clear, but the task of course differs from real shopping in some important ways, as Walkerdine notes: apart from the unrealistic prices, real shoppers do not push piles of money around to work out what their change is going to be, they buy more than one item at a time, and their cash decreases as they shop rather than being mysteriously renewed with each transaction. The "realism" of the task is maintained by the teacher's talk of buying, paying, and getting change, but this is interleaved with mathematical terms: "sum," "leaves" and "take away."

There is potential for confusion in this task, as evidenced by the fact that one child buys more than one item in the same transaction and spends all his money, while another is puzzled by a blank card. On each occasion, the teacher is forced into explicitness about the underlying mathematical task, here juxtaposing the mathematical "take" and the everyday "spend" in a reference to the problem itself:

> . . . it says start with ten pence every time doesn't it so you can work out again ten pence take . . . spend eight pence and work out how much you'll have left. (Walkerdine, 1988, p. 157)

Here the metaphor is laid bare:

Child: Miss if you get nothing on the card what do you do?
Teacher: What do you think? What do you buy?
Child: You don't write nothing.
Teacher: So you have to write nothing
. . .
Teacher: . . . so when you go to the shop you buy nothing.
Teacher: So how much money will you have for change?

(Walkerdine 1988, pp. 152–3)

Despite the potential for confusion, the majority of the children recognize the real objectives of the lesson to the extent of ignoring the shopping analogy when doing the calculations. But embedding new concepts in contexts which children are already familiar with may fudge the issue for some, as it does with Julie in my example above. As I will show in Chapter 7, the role of everyday mathematics in instruction as support for access to mathematics is not straightforward, and it remains the case that some children gain access to the discourse of mathematics while others do not. What differentiates their experiences is a major focus of this book.

Classroom Discourse and the Development of Participant Identities

As Durkin and Shire (1991) remind us, "it is worth bearing in mind that most children are very good at learning and using language . . . while very few children take so readily to mathematics" (p. 4); clearly, gaining access to mathematics is not merely a question of learning a new language. The other side of any consideration of mathematical discourse is that mathematical education takes place *in* language: educators talk to or write for pupils, while pupils make meanings which teachers respond to in return. We have already seen that mathematical texts can create differential access to mathematics, but this potential for difference is also a feature of the wider community in which mathematics teaching and learning takes place, in terms of differential patterns of classroom interaction.

That classrooms are characterized by distinctive patterns of interaction is a well-established observation. Bernstein's early focus on the relationship between educational success and social class in the UK (Bernstein, 1971–75) and his identification of the way in which middle-class children were sensitized at home to the way in which meanings are constructed within schooling drew attention to the force of particular features of pedagogic discourse in the production of "educational knowledge," explored in more detail in his later work (e.g. Bernstein, 1990). In the USA, Heath's (1983) ethnographic focus on the language and literacy practices of rural working-class communities identified similar issues in the move from home to school. Wells' (1987) longitudinal study in Bristol also drew on this type of analysis, suggesting, like Heath, that literacy practices at home were key to later school success. Contributing further to research on the home–school transition, Tizard and Hughes' (1984) study of four-year-old girls at nursery picked out differences in talk patterns between home and school and identified a Piagetian legacy in teaching styles which involved an emphasis on discovery learning by teachers who "see themselves less as teaching, and more as providing the children with a rich learning environment" (Tizard & Hughes, 1984, pp. 181–2). Drawing on Labov (1969), this study presented an important challenge to the assumption of working-class "linguistic deprivation"—a dominant view at the time, partly prompted by misinterpretation of Bernstein's concept of restricted and elaborated codes: Tizard and Hughes argued that in fact working-class children engaged in more cognitively demanding talk at home than at school but that their lack of familiarity with school discourses and a corresponding lack of shared experiences with their teachers combined to make them appear less competent than they were.

There are a number of themes from this early work which I pursue in this book. Like Wells, one of my major concerns is with the role of talk in learning, and the extent to which learners are able "to make knowledge their own" (Wells, 1987, pp. 103–24), and so develop an identity of participant in mathematics. This issue is perhaps most clearly articulated in Edwards and Mercer's (1987) neo-Vygotskian analysis of language as a tool for learning. Drawing primarily on Vygotsky (1978), but also on Barnes (1976), Barnes and Todd (1977), Dillon (1982), and Wood (1986), they identify the implicit "ground rules" of educational discourse which underlie teacher–learner interaction. These ground rules enable teachers to control the topic of discussion and the language that is used in a process of "guided construction of knowledge" (Mercer, 1995) which re-casts what is said and done in the classroom in the way that the teacher intends. The success of this process depends on the establishment of a joint frame of reference between teacher and students which draws on the immediate material context and on shared prior knowledge: in other words, teachers may be seen as working within each individual learner's Zone of Proximal Development, assisting their performance (Tharp & Gallimore, 1991), and appropriating, modifying and feeding back pupils' ideas to them in a process of scaffolding (Mercer, 1994; Wood et al., 1976). Thus Edwards and Mercer (1987, pp. 62–3) describe education as:

> a communicative process that consists largely in the growth of shared mental contexts and terms of reference through which the various discourses of education (the various "subjects" and their associated academic abilities) come to be intelligible to those who use them.

They present a detailed analysis of how classroom discourse works to build an understanding of new concepts via a collection of linguistic devices used by the teacher to guide the course of the discussion: these include responses which confirm, reject, repeat, elaborate, reformulate, paraphrase or reconstruct pupil contributions, references to implicit, presupposed and "joint" knowledge, clues of gesture and intonation to denote significance, and so on. However, like others before them, Edwards and Mercer also identified an "ideology of teaching" based on a strong discovery learning principle in the UK, rooted in the child-centered progressive education movement which began in the 1960s and was endorsed by the influential Plowden report (Plowden, 1967). Drawing explicitly on Piaget's work (Inhelder & Piaget, 1958), Plowden suggests that the ideal teacher "leads from behind," spotting readiness and providing appropriate environments which will stimulate individual cognitive advance. But, Edwards and Mercer argued, teachers' persistent questioning in order to "elicit" knowledge and promote discovery learning, coupled with the need to cover a demanding curriculum with large classes, can in fact result in merely ritual or procedural knowledge which is "embedded in the paraphernalia of the lesson, without any grasp of what it was all really about" (Edwards & Mercer, 1987, p. 99), as opposed to principled, transferable knowledge—"understanding the issues and concepts, and their relationship to the activities" (p. 99).

Edwards and Mercer's analysis, followed by Mercer's later work (Mercer, 1995, 2000), that of the National Oracy Project in the UK (a two-year initiative designed to promote oracy in classrooms; see Norman, 1992) and work on mathematics classroom cultures such as Kawanaka and Stigler's (1999) survey of teachers' questions in Japan, Germany and the USA, suggests that school cultures frequently do not foster "dialogic inquiry" (Wells, 1999). They are instead dominated by the three-part exchange of teacher initiation, student response, and teacher evaluation or follow-up (IRE/IRF) (Cazden, 2001; Sinclair & Coulthard, 1975). The power imbalance which this brings, as identified in particular by Barnes (1976) in his analysis of teacher roles in secondary schools and by Wells (1992) in his analysis of questioning and the role of pupil expertise, jeopardizes the production of shared frames of reference, forcing pupils into guesswork rather than genuine engagement (see MacLure & French, 1981) and undermining opportunities for dialogue (Myhill & Dunkin, 2005). This situation may be exacerbated by education policy: recent evidence from the UK (Hardman, Smith, & Wall, 2003) suggests that the standards and target-setting agenda of the National Literacy Hour in primary schools militates against enabling pupils to explore ideas, fostering instead a teaching style which results in the positioning of pupils "as subjects of the institutional order and its principles of classification, rather than as learners of the subject matter" (Chouliaraki, 1998, p. 8).

The assessment context seems to be highly relevant to what happens in mathematics classrooms in terms of teaching and pupil participation. Kawanaka and Stigler's (1999) analysis of eighth-grade mathematics teaching in Germany, Japan and the United States demonstrates considerable differences in teaching styles and the use of higher order questions between these three countries. Most important for my concerns here are their observations on the control of knowledge and the nature of evaluation in the classrooms they observed. Crucially it seems, both German and U.S. classrooms were highly evaluative in their use of higher order questions, whereas Japanese classrooms were far less so: although German teachers asked students what they noticed about a situation and U.S. teachers asked similarly open questions which asked for reasons and principles, both were highly controlling of the knowledge being constructed in the classroom and correspondingly evaluative in their responses. Thus Kawanaka and Stigler report that:

> It seems that the U.S. teachers used Describe/explain questions for assessment purposes more often than to elicit students' ideas and thoughts. Like the German teachers, the U.S. teachers often evaluated students' responses as right or wrong, and they rarely asked students to reflect on each other's responses. (pp. 272–3)

The fact that their higher order questions were enmeshed in assessment presents an important difference from the Japanese teachers' approach, which displayed several important features, including an emphasis on "divergent problem solving"—i.e. inviting students to work out their own ways to solve a novel problem, emphasizing the means of solution rather than the solution itself. During these episodes, teachers gave students time to think, encouraged multiple methods of solution, and treated the situation as a joint endeavor, steering away from individual learner–teacher dialogue:

> ... student responses were addressed to the entire class and not just to the teacher, and teachers tended not to evaluate student responses. Instead, the teachers encouraged the rest of the class to do the evaluation. These chains of activities ... suggest that the Japanese teachers used Describe/explain questions not only to check students' knowledge of introduced rules and procedures but also to guide students to communicate their individual thoughts and ideas to the whole class. (pp. 275–6)

Although Kawanaka and Stigler suggest that the teachers in their study simply ask higher order questions to match different pedagogic goals, the question remains as to what drives those goals and, of course, what difference it makes. The emphasis on assessment and on teacher authority in the U.S. and German contexts seems significant however, and Kawanaka and Stigler conclude that pedagogy needs to encourage students to "go beyond merely retrieving isolated concepts or processing pieces of information" (p. 278).

Ironically, the Plowden report observed that teachers needed to be more open to children's exploration in progressive teaching and consequently change their own relationship to mathematical knowledge: "this sort of approach demands a considerably greater knowledge of mathematics or rather degree of mathematical understanding in the teachers than the traditional one. If the children have to think harder, so do the teachers" (Plowden, 1967, para. 655). Forty years on, Barnes (2008, np) suggests that, " . . . the way teachers interact with their pupils is closely linked to their preconceptions about the nature of the knowledge that they are teaching. If they see their role as simply the transmission of authoritative knowledge they are less likely to give their pupils the opportunity to explore new ideas." Veel (1999, pp. 187–9) makes a similar connection between the "strongly classified" (Bernstein, 1996) nature of school mathematics and the particular dominance of teacher talk in its classrooms, drawing attention to Martin's (1992) analysis of the teacher's self-positioning as "primary knower" versus the learner's "secondary knower" status. This ascription has important consequences for the student's relationship with respect to mathematical knowledge:

> The long-term result . . . is that students are rarely given the opportunity to occupy the role of the "knower" or "producers" of mathematical knowledge. They do not get the chance to construct extended and grammatically complex text . . . in which generalized mathematical ideas are combined with measurable quantities in order to spin a rich web of mathematical knowledge. (Veel, 1999, p. 192)

This positioning as secondary knower, and the inter-relationship between beliefs about mathematics and its learning, and the nature of classroom practices, is a major focus in this book. One further consideration in this respect relates to the development of individual agency and individual trajectories in mathematics—while work in the neo-Vygotskian tradition presents an analysis of the "guided construction of knowledge" (Mercer, 1995) on a broad scale, there are further questions to ask about individual differences in engagement over time, the more long-term effects of differential access to a shared mental context, and the different responses that learners make to similar situations in terms of their *self-* positioning as mathematics learners.

As Kawanaka and Stigler show, there are important differences between countries in mathematics teaching. There is also considerable variation in how mathematics learning is experienced within the same area, as in this description of two different London schools from Boaler (1997a):

> [At Amber Hill] . . . students are taught mathematical problems but they are not encouraged to locate these within wider mathematical perspectives, they can only develop a "procedural" knowledge . . . and this knowledge is extremely limited in its applicability. . . . [By contrast] the absence of set procedures and algorithms from the Phoenix Park students'

knowledge may have given them the freedom to interpret situations and develop meaning from them. . . . the introduction of mathematical ideas as part of meaningful activities seemed to enable the Phoenix Park students to develop an inherent understanding of the meaning of the procedures they used and they learned to regard them as adaptable and flexible. (pp. 107–8)

Central to the contrast between Boaler's two schools is the extent to which pupils are able to develop agency as learners. Wells and Arauz' (2006) analysis of the impact of an emphasis on dialogic enquiry with a group of teachers suggests that it is possible to develop a pedagogic approach which fosters learner agency over time. The crux of their argument is that the IRF exchange need not in itself be problematic:

At a macrolevel, the IRF structure can be seen to aptly characterize the teacher's major responsibilities. As the participant primarily responsible for the classroom community's engagement with the prescribed curriculum, the teacher selects and prepares curriculum units and launches them in ways designed to provide appropriate challenges for each student member. This is the *initiation.* Students, in turn, are expected to *respond* by taking up some of the challenges presented and by attempting, either alone or in collaboration with others, to go beyond their current understanding or level of skilled performance. The teacher's *follow-up* then consists in responding to the students' attempts by providing assistance in a manner that jointly creates a zone of proximal development that enables them to "go beyond themselves" (Vygotsky, 1987) in relation to the challenges that they have taken up and to which they are personally committed. Viewed from this perspective, the IRF discourse genre—when appropriately used—can be seen as playing out at the more microlevel of the coconstruction of meaning the same fundamental responsibilities as are involved in organizing the more macrolevel activities in which the students are engaged. (Wells & Arauz, 2006, p. 421)

There are a number of caveats in this summing-up which I will pursue in the latter part of this book. If learners are to "go beyond themselves" to connect ideas and formulate them in a mathematical but critical way (Barnes & Sheeran, 1992), they need to perceive themselves as having some agency and as having a certain relationship to the object of their learning—i.e., mathematics. At the same time, however, teachers have a role to play as Wells and Arauz point out in terms of a legitimate, epistemic, authority in the classroom. In the next section I will begin to develop a theoretical framework for how these two aspects of teaching and learning mathematics go together.

Theorizing Mathematical Literacy

In order to gain access to mathematical literacy, learners need to develop identities of participation in the practice of mathematics in terms of ownership of its modes of meaning-making. The nature of classroom discourse is key to this development: it sets the parameters within which learning takes place, making available a range of identities from passive receiver of procedural knowledge to active participant and meaning maker. These identities delineate particular relationships to knowledge which are visible in terms of learners' understandings of the discourses of mathematics and their self-positionings with respect to its practices. As a means of exploring the development and nature of these relationships, I will draw in this book on two major theoretical strands: the theory of personal epistemologies provides tools for describing learners' perceptions of mathematics and of themselves as mathematics learners, and socio-cultural theory provides a means of capturing the dynamic of their membership of particular communities of practice. Underlying and uniting both of these stances is a focus on identity in terms of how we are positioned and how we position ourselves as mathematics learners with reference to the available cultural models.

Personal Epistemologies

Conceptions of knowledge and learning have been extensively researched since Perry's (1970) study of Harvard college students' beliefs and their development. His scheme of nine different "positions" on knowledge appears stage-like in its hierarchical move from "duality" (knowledge is true or false and based in authority) through "multiplicity" (knowledge is relative and teachers are experts rather than authorities) to "commitment" (one's own point of view is necessary for an evolving sense of identity). However, Perry rejected the concept of stages on the grounds that this presented too static a picture of the learner, and also that it failed to capture his finding that a learner could take up several positions at the same time, development being an evolving series of transitions between various positions. His theory has been seen as the forerunner of standpoint theory (Perry, 1999, p. xii) in his claim that we each interpret the world from a different position according to our particular experience, including experience related to gender, race, culture, and socioeconomic class; thus it is seen as complementary to Belenky, Clinchy, Goldberger, and Tarule's (1986) *Women's ways of knowing*, of which more in Chapter 7 (Perry, 1999, p. xix). More generally, however, Perry's approach is useful for my current purposes because it links beliefs about mathematics with individual experience, both in terms of the effect of our experiences on our beliefs, but also with respect to the effect of those beliefs on our future self-positioning in terms of our take-up of the identities which are made available within a particular social practice.

Perry's work has influenced a range of more recent research on "personal epistemologies" (see Hofer & Pintrich, 2002) and their relation to educational achievement. While some accounts follow Perry's model in making explicit assumptions about what constitutes "epistemological maturity" (Baxter Magolda, 2002; King & Kitchener, 1994; Kuhn & Weinstock, 2002), others, such as Schommer-Aikins (2002) and Schommer (1994), question the notion of "more sophisticated" epistemologies, suggesting instead that individuals have non-hierarchical systems of independently developing beliefs. Many developmental theories tend to assume a Piagetian equilibration model of change, but recent work has also considered the role of affect, motivation and context (Pintrich, Marx, & Boyle, 1993) and "epistemic doubt" (Bendixen, 2002). An important feature of both developmental and multi-dimensional models is their frequent assumption that personal epistemologies are stable, trait-like theories composed of unitary concepts which are consistent across diverse contexts. For instance, Hofer and Pintrich (1997a) propose four general epistemological categories of beliefs about (1) the certainty of knowledge, (2) knowledge as a collection of isolated facts rather than interrelated facts, (3) the sources of knowledge, and (4) how knowledge claims are justified. There are exceptions to this stance, however: building on Morrison and Collins' (1996) emphasis on the need to recognize the existence of different ways of knowing in different communities of practice, Hammer and Elby (2002) argue that epistemologies are, rather, sensitive to context and domain, and are more usefully characterized in terms of more primitive, naïve epistemological *resources* for understanding the nature and sources of knowledge, epistemic activities and forms, and epistemological stances. This seems to be a useful perspective from which to approach identity as a mathematics learner.

Closely aligned to Hammer and Elby's critique is work which focuses on discipline-specific epistemological beliefs (see, for example, Alexander & Dochy, 1995; Hofer, 2000; Hofer & Pintrich, 1997b; Lonka & Lindblom-Ylanne, 1996). Investigation of specific discipline-related beliefs has tended to concentrate on science, but it has also examined the case of mathematics, where investigation has identified a cluster of beliefs which have far-reaching consequences. De Corte, Op't Eynde, and Verschaffel (2002) sum these up:

> Mathematics is associated with certainty, and with being able to give quickly the correct answer; doing mathematics corresponds to following rules prescribed by the teacher; knowing math means being able to recall and use the correct rule when asked by the teacher; and an answer to a mathematical question or problem becomes true when it is approved by the authority of the teacher. (De Corte et al., 2002, p. 305)

This is a pervasive image of mathematics, and its importance lies in the way in which it is reflected in pedagogy, influencing expectations on the part of both pupils and teachers about the nature of knowledge and authority. As we have already seen, beliefs about mathematics are amplified in classroom discourse and are lived out in the strongly framed IRF exchange, classifying learners in terms of grading and performance. As a subject it is unquestioningly taught in "ability" groups in the vast majority of UK schools, at both primary and secondary levels (Ireson & Hallam, 2001), where ability is defined at institution level in terms of performance under test conditions (the ubiquitous "SATs"). At student level, "being good at maths" is marked by the appearance of effortless performance, speed in producing answers, and, in the late secondary years, membership of the select sub-group which takes public examinations earlier than the norm (Boaler, 1997b; Boaler, Wiliam, & Brown, 2000). It tends to be seen as an individual pursuit, as more suited to boys in terms of content (Mendick, 2005a, 2005b, 2006; Paechter, 2001) and pedagogy (Becker, 1995), and as a matter of memory, not creativity. As Schoenfeld (1992, p. 359) says, the common perception which sustains and is sustained by traditional teaching is that:

Mathematics problems have one and only one right answer.

There is only one correct way to solve any mathematics problem—usually the rule the teacher has most recently demonstrated to the class.

Mathematics is a solitary activity, done by individuals in isolation.

The mathematics learned in school has little or nothing to do with the real world.

Formal proof is irrelevant to processes of discovery or invention.

These examples are highly specific to the practice of mathematics itself, but there are further, more wide-ranging discourses about the place of mathematics in our world, captured here by Mendick (2006, pp. 17–18), who notes that mathematics is frequently depicted as any or all of the following:

- A route to economic and personal power within advanced capitalism.
- A key skill, a source of knowledge necessary for the successful negotiation of life in a scientifically and technologically sophisticated society, and thus as a source of personal power.
- A process for discovering a body of pre-existent truths.
- The ultimate form of rational thought and so a proof of intelligence.
- Associated with forms of cultural deviance where, particularly in the media, mathematicians are depicted as "nerds," a species apart.
- A skill linked to a particular portion of the human genome.

These are very powerful discourses, and much recent research has focused on their role in learning and participation in mathematics, including beliefs about the nature of mathematics itself (Lampert, 1990; Schoenfeld, 1988, 1989, 1992); about mathematics teaching and the classroom context (Boaler, 1999; de Abreu, Bishop, & Pompeu, 1997); and—most important in the current context—beliefs about the *self* in relation to mathematics learning. These studies vary in focus, including motivational patterns and self-efficacy (Kloosterman, 1996; Kloosterman & Coughan, 1994); emotional investments in mathematics (Bibby, 2002; Hannula, 2002); and gendered approaches to learning, agency and identity-formation (Becker, 1995; Boaler & Greeno, 2000). A recent extensive overview of images of mathematics and mathematicians in popular culture among 15- and 16-year-old school students and undergraduates enrolled on mathematics, humanities and social science programs in the UK by Mendick, Epstein, and Moreau (forthcoming) adds to this picture. Their respondents reported what they recognized to be clichéd images of mathematicians as old, white, middle-class men who lack social skills, have no personal life outside of mathematics and are obsessional:

> These discourses about mathematics and mathematicians are related through the idea that "normal" people engage with everyday mathematics, while other and othered mathematicians engage with its esoteric forms. Participants showed a critical awareness that the images were clichés and often both used them and distanced themselves from them. However, they were unable to produce alternative ideas about mathematics and mathematicians because of the lack of these available within their experiences of school mathematics and popular culture. (p. 20)

Picker and Berry's (2000) research on images of mathematicians confirms the widespread nature of the association of mathematics with difficulty, boredom and masculinity: their participants in Finland, Sweden, Romania, the UK and the USA most frequently drew "a white, middle aged, balding or wild-haired man" (p. 89). Crucially, these images also convey expectations for pupils about their own identities in the mathematics classroom, and the range of positions that are available to them.

Communities of Practice: Mathematics and Identity

The discourses and practices of mathematics create, and are created by, a number of intersecting communities of practice in which learners position themselves and are in turn positioned—by teachers, peers, family and schools. Lave and Wenger's (1991) account of the functioning of communities of practice focuses primarily on how the novice starts out on the periphery of a practice, moving towards greater participation as they learn through their experience and the guidance of experts. This is not an inevitable outcome, however, and Wenger's (1998) expansion of this idea involves an explication of the potential for marginalization of the individual so that they remain on the periphery, excluded from the practice. Identity is central to this process, and an understanding of identity as both part of the dynamic of learning and also its outcome is a means of explaining how individuals take up certain positions with respect to mathematics.

Wenger (1998, p. 151) defines identity as "a layering of events of participation and reification by which our experience and its social interpretation inform each other. As we encounter our effects on the world and develop our relations with others, these layers build upon each other to produce our identity as a very complex interweaving of participative experience and reificative projections." Thus, because we come into contact with several practices, some of which we are full members of and some of which we are not, "we know who we are by what is familiar, understandable, usable, negotiable; we know who we are not by what is foreign, opaque, unwieldy, unproductive" (p. 153). Identity is therefore cumulative: it is built over time as we participate (or not) in a community of practice, and the nature of our participation and our location of ourselves is interpreted in terms of the values, assumptions and rules of engagement and communication of the practice. This appears to fit educational experience very well—a learner's actions in the classroom are interpreted and those interpretations in turn influence further actions within a network of relationships which include teachers and pupils. Identity in this sense is not fixed, of course, and individuals move in and out of practices on a number of trajectories, sometimes on the way to full participation, sometimes as an insider participating in the evolution of a practice, sometimes as a peripheral participant, sometimes as one who spans the boundaries between practices, and sometimes on the way *out* of a practice (pp. 154–5). Thus Wenger suggests more than one "mode of belonging" (p. 173): identity has multiple aspects of imagination, alignment and engagement.

Engagement in a practice is characterized by active negotiation of meaning within it—our appropriation of those meanings is what enables an identity of participation (p. 202). It represents an embeddedness within the practice and from that point of view may be seen as an ideal in education terms. However, the notion of trajectories underlines the fact that identity is not static and is not solely composed of community membership and its associated competencies. Individuals are equally defined in terms of their non-participation in practices and movement between or within them. Thus, in addition to engagement, Wenger invokes imagination and alignment as further aspects of identity. Imagination involves standing back from direct engagement and an awareness of actions as part of historical patterns and potential future developments, of others' perspectives and of other possible meanings. It is "a process of expanding our self by transcending our time and space and creating new images of the world and ourselves" (p. 176) and so involves a positioning of self with respect to our own and other practices. Importantly, it enables an identity of learner. In essence, this aspect of identity is one of self-awareness and self-reflection, a positioning of self within the social nexus of practices. Alignment, on the other hand, emphasizes common patterns of action. Wenger's examples of scientific method and educational standards which "propose broad systems of styles and discourses through which we can belong" (p. 180) are particularly relevant here and clearly apply to the learning of mathematics. Thus we position ourselves through adherence to these more global practices and identify ourselves in accordance with them. Thus these three modes of belonging are complementary, and indeed Wenger argues that "they work best in combination" (p. 187).

However, each has trade-offs. While alignment has a positive coordinating aspect, it also has an element of control, and as Wenger points out, it involves power—either to align oneself or in the demand for alignment from others: although it is important in maintaining a practice, an identity of non-participation can be generated by demands for alignment when the ownership of meaning is not shared. Similarly, imagination can be used positively to assess, control, and even resist the negativity of alignment through its grasp of the positioning of self, but imagination can also wrongly assume that stereotypes hold true or that others must see things as we do. Thus Wenger describes it as "a delicate act of identity . . . [which] . . . runs the risk of losing touch with the sense of social efficacy by which our experience of the world can be interpreted as competence" (p. 178). Engagement too needs to be counterbalanced by the positive creativity of imagination; it can become so narrow as to close down our ability to see from other perspectives or to stand back and reflect on the deeply embedded practices of which we are a part. Wenger thus describes a situation of stagnation in which "competence can become so transparent, locally ingrained, and socially efficacious that it becomes insular: nothing else, no other viewpoint, can even register" (p. 175).

Thus our relationship to a practice is more complex than might at first appear to be the case. It is the mix of modes of belonging and their related identities of participation and non-participation that constitutes, and is constituted by, the extent to which individuals identify with a practice and are able to control and negotiate meanings within it. Within the mode of engagement, positive outcomes for the learner are to be able to "appropriate the meanings of a community and develop an identity of participation" (p. 202); but it is also possible to exclude learners from the negotiation of meaning so that "members whose contributions are never adopted develop an identity of non-participation that progressively marginalizes them" (p. 203)—we need an ongoing identity of participation in order to learn. Within the mode of imagination, lack of access to a practice can become marginalization as the individual assumes that meanings "belong elsewhere" and will never be theirs to access. Finally, within the mode of alignment, while initial guidance and modeling introduces the learner to the possibilities of a practice, lack of ownership generates and is generated by compliance and an emphasis on procedures (p. 205). Thus we can see a source for what Edwards and Mercer (1987) and Mercer (1995) call "ritual knowledge" as a product of teaching in which learners have no control and simply seek to supply teachers with "right answers," relying on the "paraphernalia of the lesson" as a prop which, when removed, leaves the learner without the ability to apply knowledge in new situations:

> ... literal compliance can be efficient, since it does not require the complex processes of negotiation by which ownership of meaning can be shared. But for the same reason, it is brittle in that it makes alignment dependent on an environment that is specifically organized, conforming, and free of unforeseen situations. Such lack of negotiability can only engender either strict alignment in terms of the reification or no alignment at all, which results in an inability to adapt to new circumstances, a lack of flexibility, and a propensity to breakdowns. (Wenger, 1998, p. 206)

Thus alignment has a part to play in learning to be a participant in a community of practice. Without it, learners meaninglessly exercise imagination in a social vacuum. But ownership of meaning and the possibility of creating new meanings is necessary if individuals are not to become marginalized: they must retain the identity of the peripheral participant on an inward trajectory, rather than an identity of marginalization.

Wenger's framework of identity thus provides us with tools for exploring an individual's relationship with a community of practice. In Chapters 3, 4 and 5 in particular I will apply it to what mathematics students say about their experiences of learning in secondary school and at university as a means of analyzing their participation—or engagement—in the central meaning-making practices of mathematics. However, Wenger's account of the dual nature of identity reminds us that "Identification without negotiability is powerlessness— vulnerability, narrowness, marginality. Conversely, negotiability without identification is empty—it is meaningless power, freedom as isolation and cynicism" (p. 208). I explore the implications of this particular observation fully in Chapter 8.

Identity and Mathematics—What and Who This Book is About

Identity is central to any socio-cultural account of learning. As we have seen, it is central to students' beliefs about themselves as learners and as potential mathematicians. Thus understanding identity in terms of a learner's place in a community of practice, or their *perceived* place in it, seems to be the key to understanding exclusion from, and inclusion in, mathematics—whether or not we see ourselves as a member of the group of people who like and can do mathematics or, more importantly, *potentially* a member of that group, determines in great part our actions and beliefs as mathematics learners. This is most visible in formal learning contexts where learners are subject to institutional structures which impose categorizations on them as good at or not good at mathematics via assessment, curriculum, classroom discourse and physical groupings. Learners, drawing on their experiences of schooling and their wider identities of gender, ethnicity and class in turn construct corresponding mathematics identities as they negotiate the educational system and the discursive positionings which it makes available to them. Bearing in mind the research overviewed already in this chapter, we can see that identity has a major part to play not only in our understanding of pupils' perceptions of their mathematics ability but also in their sense of ownership and participation in mathematics as a creative activity—their personal epistemologies of mathematics.

In this book I explore the processes involved in promoting a mathematical literacy which has its basis in the development of an identity of inclusion and all that that involves: particular beliefs about oneself as a learner and about the nature of mathematics, an identity of engagement in mathematics and a perception of oneself as a potential creator of, or participant in, mathematics. I base this exploration on data drawn from a number of different case studies: from recordings of mathematics lessons spanning a period of 5 months in a Year 5 (aged 9–10) class of 28 children collected by my colleague Laura Black[2]; from interviews with 27 secondary school children in Years 7, 9 and 10 (aged 11–12, 13–14 and 14–15); and from focus groups and interviews with 27 mathematics undergraduates. I refer to these groups throughout the book, although particular groups feature in particular chapters. I introduce them briefly here; readers who are unfamiliar with the education system in England and Wales will find more information on how the institutions in which these data were gathered fit in to the broader context in Appendix A, while details of the samples, data and analysis can be found in Appendix B.

The Primary School Children

The children attend a school based in a large town in the north-west of England. It is one of two in the town which has a concentration of British Asian children and it is a larger than average school in the town. The proportion of pupils eligible for free school meals (19%) is in line with the national average, and the number of pupils learning English as an additional language (13%) is higher than average. Economically and socially, the area served by the school is generally similar to the national picture, and attainment on entry to the school is broadly average. The classroom observed was a Year 5 (aged 9–10) top set maths class and consisted of 12 girls and 16 boys; four pupils were from minority ethnic groups (one Chinese girl, one African-Caribbean boy and two Asian boys). I have built on Black's initial analysis of the children into talk-type groupings (see Black, 2004c); the children and their group membership are listed in Table 1 of the Appendix.

The Secondary School Students

The data from this group are drawn from interviews with 27 girls and boys in Years 7, 9 and 10 (ages 11–12, 13–14 and 14–15) attending a mixed 11–16 comprehensive school in the north-west of England. The school teaches students in heterogeneous "mixed ability" groups in Year 7, their first year at the school, but groups them for mathematics lessons from the beginning of Year 8 in accordance with Year 7 performance scores. The students' teaching set membership is listed in Table 2 in the Appendix. The school has very few ethnic minority students, and all pupils have English as their first language. The school draws on a mainly rural population; though mixed, the area is generally socially disadvantaged. The attainment of pupils on entry is judged as average by government standards.

The Undergraduates

The undergraduate data were collected in three locations: in interviews with first-year undergraduate mathematics students at Bradley, an English university with a strong research culture, in joint interviews with second- and third-year undergraduates at Farnden, an English teaching-led university, and in focus groups with second year students at Middleton, an English research-led university. Bradley University demands high entry qualifications and operates a choice system whereby students enter the University to study a particular subject, but may change to a different degree program at the end of their first year, providing they have studied the prerequisite first-year courses: this is possible since students can study within up to three different discipline areas in their first year. Middleton also requires high entry qualifications, although these are slightly lower than Bradley's. Farnden requires much lower qualifications for entry. Both Middleton and Farnden operate more conventional degree programs, with no choice process as at Bradley. However, it is possible to transfer within the mathematics discipline between pure and more applied programs. Bradley first-year students' second-year choices and Farnden and Middleton degree schemes are listed in Tables 3 and 4 in the Appendix.

In the next chapter, I will begin to look in detail at the nature of mathematics classroom cultures and the means by which they produce and reproduce particular identities of inclusion and exclusion, drawing on previous research on classroom cultures in the compulsory years of schooling and introducing some of the primary school data in order to set the scene for a detailed analysis of the secondary school students' stories in Chapter 3.

2
Experiencing Mathematics
Building Classroom Cultures and Mathematics Histories

What are the origins of our relationships with mathematics? We know that many learners shy away from mathematics, seeing it as an activity in which they fail, or fear to fail, while others associate it with excitement or comfort—mathematics appears to engender strong emotions (Black, Mendick, & Solomon, forthcoming; Breen, 2000; Buxton, 1981; Evans, 2000; Ma, 2003; Nimier, 1993)—it seems that very few people are indifferent to it, having a story to tell which may be positive or negative: they are unlikely to have no story at all. In this chapter, I will examine the crucial elements in the classroom community of practice which contribute to this range of relationships and the potential for development of identities of participation for some learners but not for others. Drawing initially on primary school data to illustrate the available range of identity choices that learners have in this environment, I will suggest that the ways in which students are able to engage with mathematics are shaped by the Discourses (in Gee's 1996, 1999 sense) of which they are a part and, further, that these Discourses intersect in complex ways which may have far-reaching consequences.

The Mathematics Classroom Culture

Although every classroom culture is unique, with its own particular dynamic and patterns, it is also true to say that classrooms—in the compulsory school years at least—have particular and important characteristics, as Lemke (2000, p. 278) points out:

> Are there emergent processes and patterns in classrooms? I think every teacher and student knows that there are. There are new routines that emerge, new social groupings and the typical interactions that sustain them, class in-jokes, informal rituals, typical sayings and phrasings, favorite word usages with special meanings, and so forth. These in turn can become the raw material for more complex new patterns unique to the classroom, and they certainly constrain the probabilities of actions and utterances that would invoke these special meanings or contribute

positively or negatively to social relationships. A classroom, and indeed every human community, is an individual at its own scale of organization. It has a unique historical trajectory, a unique development through time. But like every such individual on every scale, it is also in some respects typical of its kind. That typicality reflects its participation in still larger-scale, longer-term, more slowly changing processes that shape not only its development but also that of others of its type.

There has been much research on what is typical in mathematics classrooms, focusing on pedagogy, curriculum and assessment and I will review some of this here. There is rather less on the emergent social groupings and interaction patterns that Lemke talks about in the first part of this quotation, although these are implicit in research on gender and "ability" groups in mathematics classes and the ways in which individual students position themselves within them. This research draws not only on patternings within individual classrooms but also on wider and more long-term social structures involving gender, class and ethnicity. The issue, as Lemke argues, is to understand how longer-term processes interact with shorter-term ones, and one way in which we can do this is to study how the individual constructs their identity within a social network—this individual acts as a "boundary object" (Star & Griesemer, 1989)—something which (or in this case, some*one who*) acts as a bridge between different situations, playing a different role in each:

> Students interact with one another and with the other available semiotic objects in various intersecting activities, and these activities are recognizable and repeatable and usually repeated. In this participation we learn to do differently and to be different. We engage with a person or an artifact in a particular way, typical of that activity, and now the system in which our *persona* exists and functions changes. Dynamically, we are what we do, and we are now creating ourselves as personae in interaction with new others and artifacts. (p. 285)

So, we need to keep in mind Lemke's starting problem, which is to explain "How do *moments* add up to *lives?*" (p. 273), or, to be more precise: how do *mathematical* moments add up to *mathematical* lives or *histories*, as they are lived, constructed, narrated or even foretold?

Typical Patterns in Mathematics Classrooms—Initial Issues in Pedagogy and Culture

As the research I reviewed in Chapter 1 suggests, our relationships with mathematics have their roots in the pre-school years, via two major strands of experience: the informal numeracy practices that we participate in, and our early experiences of pedagogic discourse. In terms of our enduring relationships with classroom mathematics, it is the latter which is the most important, since it is this experience which will have an effect on our understanding of and engagement with mathematics as it is presented within the discourse of schooling. In so far as teaching builds on what we already know, then our understanding of everyday experience as it is filtered through this discourse is dependent on our facility with it and our ability to separate out the schooled from the informal.

However, Bernstein and many other researchers since have argued that there is differential access to this discourse as a result of socio-economic differences in children's home lives; if this is indeed the case, then there will be consequences for access to school learning, and particularly mathematics as a strongly classified and framed subject, as Lerman and Zevenbergen (2004) suggest. In her study of mother–child dyads, Hasan (2002) has attempted to provide empirical evidence for the existence of social class-based differences in pre-school exposure to the kind of talk which enables a sensitivity to pedagogic discourse and hence prepares children for schooling. Following Bernstein's focus on workplace positioning, she distinguishes between families in terms of their membership of "dominating and dominated social groups," defined in terms of "the degree of control on the workplace environment" (p. 540), arguing that mothers from families having greater workplace control (i.e. middle-class professional) engage in talk with their pre-school children which significantly differs from the mother–child talk in less privileged (i.e. working-class manual) families. In her research, middle-class mothers tended to use an "informative mode"—"*an explicit mode of attempting to inculcate some concept, irrespective of the domain within which the concept is located*" (p. 542, italics original)— whereas the working-class mothers used a "formative mode": "rather than instilling particular pieces of information, *it functions as a way of setting up interpersonal relations between the discursive dyads*" (p. 543, italics original). Informative talk involves a deliberate and explicit attempt to transmit knowledge, whereas formative talk carries assumptions which are implicit but treated as self-evident.

The distinction lies in the degree to which talk assumes an intellectual separation between interlocutors, thus relying on explicitness for good communication: "if the personal distance between the speaker and the addressee has to be bridged discursively, then the terms in that discourse need to be carefully displayed, and information has to be detailed" (Hasan, 2002, p. 544). The distinctions that Hasan notes are not just evident in mothers' talk, but also in their children's, an observation of some importance in that it provides support for the claim that what happens at home makes a difference to what might happen in school in terms of the range of positions available to the child. Particularly pertinent for the transition to classroom discourse are the features of questions and answers: in the middle-class families, questions were more often prefaced (did you know?), related (i.e. relating ideas), and elaborated (you can't do that because . . .), while answers were more often responsive, adequate (i.e. not minimal yes/no), related, and elaborated. Their unifying characteristic is the fact that they do not assume that the speaker knows or thinks what the listener knows or thinks, whereas those in working-class mother–child dyads carried a core feature of assumption that this was the case. Contrasting the informative and formative modes, Hasan argues that the consequences are far-reaching:

> The expectations of discursive engagement that the two groups of children entertain, the principles for interactive practices that they internalise, are already being learnt at this early stage, and the learning varies systematically, in keeping with the predictions of Bernstein's code theory. (p. 546)

Thus, those children who experience informative discourse are in a position to assess what is relevant and to access the specialized discourses of school, such as school mathematics. By contrast, those who have experienced mainly formative talk are not merely excluded from specialized discourses, but take from the classroom something rather different: their expectations about the nature of engagement mean that they are attuned to the regulatory aspect of the pedagogic discourse only. Hence Hasan claims that "what we actually learn in our lifetime is typically constrained by our social location" (p. 537)—so, as Bernstein (1996, p. 32) argues,

> Many children of the marginal classes may indeed have a recognition rule, that is, they can recognize the power relations in which they are involved, and their position in them, but they may not possess the realization rule. . . . These children in school, then, will not have acquired the legitimate pedagogic code, but they will have acquired their place in the classificatory system. For these children, the experience of school is essentially an experience of the classificatory system and their place in it.

While this assessment of educational access is difficult to observe directly, evidence of differences in classroom interaction patterns can suggest ways in which some pupils, but not others, participate in the "legitimate pedagogic code" and so gain access to particular knowledges and knowledge-making practices. Black (2002a, 2004a, 2004c, 2005) presents an analysis of classroom interaction between a group of pupils and Mrs. Williams, their teacher, in a typical British primary school over a period of five months. Methodologically, her account of this exercise shows that it is possible to make sense of classroom interaction patterns on a number of levels (see in particular Black, 2004c), and to identify trajectories of participation in which pupils are positioned by the teacher, but also position themselves, within different intersubjective frames. Thus individual pupils can be seen to play consistent roles which appear to be linked to levels of cultural capital. Another way of describing these roles is in terms of Gee's

(1999) Discourses, as recognized and accepted ways of being which draw on cultural models within a particular social situation:

> A Discourse is a socially accepted association among ways of using language, other symbolic expressions, and "artifacts," of thinking, feeling, believing, valuing and acting that can be used to identify oneself as a member of a socially meaningful group or "social network," or to signal (that one is playing) a socially meaningful role. (Gee, 1999, p. 131)

For example, Phillip is a member of "Group A," a group of eight middle-class boys (from a class of twelve girls and sixteen boys) who not only experienced far more interactions with the teacher than the other twenty class members but who also experienced more "productive" (i.e., dialogic or exploratory) than "unproductive" (i.e. didactic, teacher-controlled) interactions, a distinction which draws primarily on the work of Barnes (1976) and Edwards and Mercer (1987) (see the Appendix and also Black, 2004a, for details of these categories and their application). Here we see him responding to a request for ideas on how to represent the weather graphically in an extract taken from Black (2004a, pp. 40–1):

Phillip: You could put them . . . like the Monday underneath it like that.

T: You could. You could put Monday, Tuesday, Wednesday, Thursday, Friday at the bottom of your graph. That's true. So let's assume it's going to be just like most graphs it has a vertical and a horizontal axis and at certain points it has little bits of information. And at the bottom Phillip you're suggesting in these boxes at the bottom we put Monday, Tuesday, Wednesday, Thursday, Friday. *(drawing it on the board)* We can't do Saturday and Sunday can we but we can do the following Monday, Tuesday, Wednesday, Thursday, Friday. Yeh? So what we gonna do now? How we going to . . . how we going to show the weather on Monday of last week? *(pause)* Well it was sunny. What could you do to show that?

Phillip: Put it in a colour . . . of the sun. Put yellow.

T: Good idea. So instead of using the symbols as they are, we could make them into the colour to represent that symbol. So if we're gonna use . . . if we're gonna use colours to represent sun and rain and fog and so on, what else are we going to need on our graph?

Phillip: What the temperature is.

T: No we're thinking about the colours now. Are those colours gonna mean anything to anyone apart from you? Unless you do what?

Phillip: You put a little key down the side.

T: Little key to represent the colours. So that yellow equals sunshine.

This is typical of Phillip's exchanges with Mrs. Williams, in which he shows his understanding of her pedagogic intentions and references (including her reference to past work on keys)—and in which she in turn takes his contributions seriously and persists with the dialogue even when it is in need of repair as it is when Phillip initially misunderstands what she is asking for when she is seeking the word "key." The exchange, in this sense in particular, can be said to fit with Grice's "cooperative principle" (Grice, 1975): the teacher appears to recognize that Phillip's (incorrect) contribution is from his point of view relevant (and therefore does not indicate lack of ability), but that since this is not what she intended she must make herself clearer. This repair sequence is arguably the most indicative feature of Phillip's positioning in the classroom. As Wells and Arauz (2006) suggest, it is a feature of the participants' drive towards the maintenance of intersubjectivity. Its importance is seen most clearly when this dialogue is compared with the following one involving Hasan, a boy who belongs to "Group B," the group of five children—four boys and one girl—who experienced more unproductive than productive interactions, in Black's analysis. In this extract, the children are learning about categorizing shapes in Venn diagrams:

T: So B will go right in the middle there, won't it Hasan? (*she is pointing to the centre of the Venn Diagram in the textbook*) B, Do you see why? It will go in the middle there? (*no response*) Do you see why it will go in there? Can you explain why?

Hasan: It's got five faces.

T: Pardon?

Hasan: It's got five faces. (*louder*)

T: Good, it's got five faces, what else?

(*Silence. Group A pupils have their hands up*)

T: That's one reason why, that's not the only reason why it can go in the middle, is it? What's that say there? (*teacher points to one circle in the Venn Diagram*)

Hasan: Red. (*reading from book*)

T: What does that say there? (*points to another circle in the Venn Diagram*)

Hasan: "Has at least one square face." (*reading from textbook*)

T: And that has got a square bottom hasn't it?

Hasan: Yeh.

T: an' it's red and it's also got five faces, so that's the only shape that will go in the middle, the rest you're gonna have to decide, some might go in between red and has a square face or might go in between red and has five faces, it might not belong in any of them, in which case you put the letter outside the Venn diagram.

(Black, 2004a, pp. 42–3)

This exchange contrasts most starkly with the previous one in terms of the apparent lack of willingness to cooperate on the part of both Hasan and Mrs. Williams. Hasan—knowingly perhaps—gives insufficient information, while she for her part fails to make her requirement explicit. It is worth quoting Black's analysis at length, here, since it underpins the claim that children like Hasan are disadvantaged at school in terms of their exclusion from specialized discourses and their subjection to regulatory discourse:

> ... here the teacher displays much greater control over the shape of the interaction. This is evident in the involuntary nature of Hasan's participation; he is called upon by name and strongly urged to take part in the discussion. The long process of cued elicitation involves the teacher trying to draw out the correct response from Hasan using heavily clued questions. Eventually, she resorts to forcing him to read out the name of the categories from the textbook as a semi-adequate mode of answering ... However, this is an incomplete answer and she reformulates what he has read out in order to answer her own question in the correct manner ... Hasan plays a highly passive role in the interaction and is, therefore, unable to actively seek out the teacher's understanding of the situation. (Black, 2004a, p. 43)

Hasan's apparent lack of control in this exchange is a function of both his lack of "pedagogic awareness" (Black, 2002a, pp. 85–7), and his more general failure to display the cultural capital of middle-class educational values, thus fuelling Mrs. Williams' expectations about his ability (i.e. that it is low). Pedagogic awareness in this analysis primarily encapsulates a cluster of knowledge about the shape and nature of classroom talk, and an understanding of the teacher's role as expert and the learner's role as novice/apprentice, in Lave and Wenger's (1991) sense. In Black's analysis of the pupils in her classroom, it is possession of this knowledge by Group A pupils versus its lack in Group B pupils' repertoires which is crucial in determining their educational experience:

> ... not all children start out on an equal playing field when they enter the classroom. Some are more likely to become learners than others, because their cultural background provides them with the capacity to recognise dialogic forms of behaviour as appropriate to classroom discussions, which informs teacher expectations and behaviour. (Black, 2004a, p. 51)

What this analysis suggests is that Hasan and others like him take up particular positions in the classroom which do not include involvement in episodes of exploratory talk. Neither he nor his teacher enter into the kind of negotiation mode observed by Wells and Arauz in dialogic classrooms, and indeed he is positioned by her as a particular type of learner in public exchanges and also in private conversation. In the following episode he is held up as an example to the other children.

T: And yesterday I noticed you did this wrong *(to Hasan)*. You did this wrong yesterday so just you watch *(pointing to Hasan)*. You did this wrong, I know it wasn't this sum but it was a similar sum and you said how many nines in one? Nothing. And then he said how many nines in five. *(pause)* Now that's not right, is it? Cos he's just ignored the hundred. He's just said how many nines are in a hundred and how many nines are in fifty. He's forgotten about the hundred. What should he have done in that one? *(pause)* *(Peter puts hand up)* How many nines in one? You can't. You can't share one by nine but then you say what? *(Sean puts hand up)*

Sean: Take one off the five.

T: Ooh no you're doing a subtraction sum there. *(Simon has hand up)* *(pause)* Nelson what would you do? *(pause)* with a sum like that? We've said how many nines in one? You can't. If it'd been how many . . . let's make it easier *(drawing on board)* That would've been ok cos we'd say how many twos in two? One. How many twos in four? Two. How many twos in six? Three. That's alright. But here you can't actually share the one into nine equal parts can you? So what would you do? What's the next thing we've got to do? *(pause)* *(James puts hand up and then Phillip)* Have you done this sort of sum before?

Phillip: Can you take the one off the five?

T: Well you don't take it off really. *(pause)* You move it across don't you? Or you just say to yourself "how many nines in fifteen?" Don't you?

(Black, 2002a, p. 264)

Note that, once again, Phillip's error at the end of this exchange (which is in fact a repeat of Sean's rejected contribution earlier) is instantly repaired, meeting with a very different response from that to Hasan's *non-*erroneous, albeit incomplete, contributions in the previous extract and his reported error the day before. In contrast to Hasan's public position as an example of what *not* to do, Simon, who like Phillip is a Group A pupil, is paraded as an exemplary learner in more general discussion as in the following lengthy episode of talk around the "master class" that he and selected others attend:

T: Now Simon, I don't know how many of you know, goes to some extra classes don't you?

Simon: Hmm.

T: Would you like to tell us something about that and perhaps maybe bring in some of the stuff you do Simon for us to have a look at. Would you explain to the children what you do, where you go . . .

(Some discussion about which day the classes are held on, prompted by the teacher)

Simon: Well erm we've been picked out because we're one of the . . . you know smart people in the class and the government . . .

T: *(interrupting)*: That's one way of putting I suppose. It's true though actually.

Simon: . . . wants to know if we are getting on good with it cos they want to do it . . . in the future . . .

(Some discussion about where the classes are held, prompted again by the teacher)

T: . . . Now you're not the only two *(i.e. Simon and Tim)*, who else goes?

Simon: erm Sian . . .

T: Sian and?

Simon: er . . . Dawoud

T: Dawoud and? Is that it?

Simon: er . . . no there's somebody else.

T: Daniel? Suzanne?

Simon: yeh.

T: these children have been selected yeh . . . but saying it like that makes you think "ooh a big head" but what we've had to do, we really have, we've had to look very carefully at the Richmond profiles. You know you do Richmond tests? And the various tests that you did in Year 4. And we've been asked . . . *not* our idea, let me make that clear. It was the grammar school have got this funding which means they've got a lot of money to spend for . . . is it for this year and next year?

Simon: I don't know.

T: Certainly for this year . . . to see if you can . . . they can *extend* the understanding experience . . . the academic understanding . . . the thinking in certain subjects. And those subjects are science and technology and maths. So these children go to these classes in their own time, after school on a Tuesday night and the Grammar School teachers teach them.

(Discussion about the timing of classes after school, and how the grammar school teachers give up their time for them)

T: They're guinea pigs in a way. And the idea is to see what improvements, if any, of them make in that year, given this extra special kind of teaching. Now you go to maths class that's why I asked you to tell us about some of the things you do in that class that's different to the sort of maths you do with us.

Simon: Well we are learning about you know the Egyptian maths er . . . we're doing how to do magic squares er . . .

T: Yeh could you bring some stuff in for us sometime?

(Discussion about Simon bringing in his work)

Simon: Yeh well we're allowed to take things home. I mean we don't have to do all the work.

T: No.

Simon: They just want us to learn how?

T: Yeh so it's quite free and easy. It's not a big pressure? Are you enjoying it?

Simon: Yeh.

(Black, 2002b)

I have quoted this exchange at length as it illustrates a number of issues about the perception of certain pupils and the emphasis in the school on the importance of test performance as an indicator of ability. Firstly, we see the careful positioning of Simon and his fellow master class students as members of an exemplary and explicitly named group by Mrs. Williams. She is at pains to list the members of the group, to explain the thinking behind the master classes and to emphasize the voluntary nature of teacher and pupil participation and the enjoyment and interest to be gained from the classes. Crucially, though, this is about "*ability*": she does not challenge Simon's self-positioning as "smart," although her response is complex and includes a self-distancing from his elitism ("that's one way of putting it I suppose") but also agreement ("it's true though actually"). She continues with this ambivalence, on the one hand stressing that the impetus for the master classes has come from outside ("we've been asked . . . *not* our idea, let me make that clear") and that they are experimental and (possibly) extravagantly resourced ("they've got a lot of money to spend . . . they're guinea pigs in a way"), but on the other hand endorsing the targeting of higher attainers ("these children have been selected yeh . . . but saying it like that makes you think 'ooh a big head' but what we've had to do, we really have, we've had to look very carefully at the Richmond profiles") and their suitability for a program which aims to "*extend* the understanding experience . . . the academic understanding . . . that's different to the sort of maths you do with us." In these latter statements there is a hint of a further level of ambivalence towards the National Curriculum and the approach to mathematics that it projects: one interpretation of Mrs. Williams' talk here is that she believes the standard curriculum to be in need of the kind of improvements that the Grammar School is working on and, possibly, that Richmond profiling is necessary because success in mathematics is not just about getting right answers. As Horn's (2007) analysis of two Californian schools struggling with reform agendas shows, making the shift to more inclusive pedagogy is difficult, and this is all the more so against a backdrop of audit- and standards-driven educational policy. I will return to this issue and a discussion of Mrs. Williams' pedagogic approach below.

It is clear that her intention in the discussion on the master classes is to motivate the other children, but the success of this endeavor is questionable. While Simon identifies as "smart," Nelson and Carl's private conversation suggests rejection of the master class idea in Carl's (unusual) use of Simon's family name rather than his given name, and Nelson's claim that going to the classes is "stupid":

Carl: Jones goes to Masterclass, dun he?
Nelson: What?
Carl: Jones goes to Masterclass, dun he?
Nelson: Yeh, that's stupid.

(Black, 2002b)

It is indicative that these two Group B boys reject this marker of success. In the same group, Hasan also shows a sensitivity to Simon and the other Group

A pupils' participant roles in the classroom, positioning himself and Jason as non-participants:

I: Do you think . . . you know all the children in the class . . . do they all behave
   or talk in the same way? Or do some people behave differently?
Hasan: Behave differently.
I: Differently?
Hasan: Yeh.
I: In which way?
Hasan: Like telling the different answers and stuff like that.
I: Who behaves in what . . . you know who's particular in what way they
behave? *(pause)* Who's different from each other? *(pause)* Who always
answers Mrs. Williams' questions?
Hasan: Jeremy and . . .
I: Jeremy did didn't he? He's gone now though hasn't he? Who else?
Hasan: and then Chris and Simon.
I: And is there anyone who doesn't answer?
Hasan: Jason and me.

(Black, 2002a, pp. 162–3)

Group B member Erica says something very similar:

I: Do you think, erm, everyone talks in the same way or do some people talk
   differently?
Erica: Some people talk differently.
I: In what way? How is different?
Erica: Some people describe it differently.
I: Oh right so who in particular describes things differently?
Erica: Well I know Tim does and Simon. *(Group A pupils)*
I: In what way . . . how do they describe things?
Erica: They say it in a long way.
I: In a long way? What's a long way?
Erica: They start talking about it and then they can hardly stop cos they keep
   talking about it for ages.

(Black, 2002a, p. 162)

Simon himself is confident about his ability to answer questions in class:

Simon: I usually know them *[i.e., the answers to questions]*.
*(Discussion about whether he prefers to put his hand up or be asked by the teacher
   directly)*
Simon: Yeh . . . yeh it's ok. I can usually answer the questions.
I: Yeh ok then and when you don't answer what do you usually do? When you
   don't know the answer?

Simon: If I don't know the answer . . . well . . . well I usually, I usually say nothing and then she'll and then sometimes she'll go on to somebody else. That hasn't happened often though.

(Black, 2002a, p. 248)

Privately, Mrs. Williams' constructions of the children as particular types—for example she describes Simon as "professorish" and not really needing to listen (Black, 2002a, p. 247)—suggest that assumptions about (natural) "ability" are in continual conflict with other more inclusive pedagogic aims that she may have. Here she responds to Black's use of the word "ability" as a distinguishing feature between children with respect to Hasan, in the familiar conflation of "ability" and confidence (see Hardy, 2007):

I: So you don't think there's a relationship between the sort of contribution or their participation and ability?
T: Generally yes there is, yeh. *(pause)* Yeh. Unless Hasan is really, really confident he doesn't say anything, very rarely does he offer an answer in class unless I ask him and if I ask him he still then looks a little bit reticent, "should I?—shouldn't I?" Eventually he'll give an answer.

(Black, 2002a, p. 206)

In her discussion of Nelson, she makes implicit reference to his Afro-Caribbean heritage as an explanation for his performance in terms of character while at the same time claiming that she does not see the children as different, in a way which is reminiscent of Biggs and Edwards' (1991) claim that "I treat them all the same":

I: What about Nelson though? Do you find there's differences for him? Or . . .
T: Yeh *(pause)* . . . I think, given the household that he comes from . . . lovely parents but just so laid back it's unbelievable. And Nelson has turned out great really when you think of what he could have been like, . . . it's such a sort of laid back sort of attitude. You know it's like they are . . . "he's oh, he's just like his daddy" . . . I mean he's a smashing kid so . . . and I don't think of any of them being different at all.

(Black, 2002a, pp. 201–2)

We have seen that gender appears to be a factor in the class sub-grouping. Mrs. Williams voices a concern with gender issues, but in doing so makes clear her assumptions about girls as a group in response to Black's introduction of the boys' underachievement debate:

I: Yeh I've . . . well I've read a study . . . you know there's all this stuff about boys' underachievement and homework . . .

T: In fact I'm going to a course er a week after half term on exactly that—
"underachieving boys."

I: Well I read something about it being a sort of social thing and how men don't
see their identity as, you know going home and doing homework whereas
women do because they go home from work and do the housework, you
know, so that kind of work in the household is a different . . . is something
they're much more experienced with so they'll go and do homework . . .

T: And play schools and things with their friends and that . . . boys wouldn't
would they? You see, its a bit cissy that in it?. . . I mean they seem to want to
take more care of their work, they're often tidier and more careful about their
presentation but boys catch up later on don't they?

(Black, 2002a, p. 203)

This is a familiar claim, of course, in which girls are constructed as concerned
with neatness and care whereas boys are not but nevertheless get through the
material, often more quickly (see also Walkerdine, 1998). Thus Walkerdine notes
how boys' "poor performance is both excused and turned into a good quality"
(p. 162) while girls work hard and strive for what are "feminine" qualities which
are required but at the same time de-valued:

> Girls, at the nexus of contradictory relationships between gender and
> intellectuality, struggle to achieve the femininity which is the target of
> teachers' pejorative evaluation. They often try to be nice, kind, helpful and
> attractive: precisely the characteristics that teachers publicly hold up as
> good—asking all children to work quietly or neatly, for example, while
> privately accusing the girls of doing precisely these things. (p. 162)

Mrs. Williams' positioning of the whole class revolves around a mixed
discourse of differences in ability, speed and effort. Here, she uses a public
"telling off" of Jason as an opportunity to berate the rest of the class for lagging
behind:

> T: *(to Jason)*: Mr Stroud and I have had chats about you. You're losing
> . . . you're losing that little extra something you had at the beginning of
> the year, you're not trying as hard and that worries us because you were
> beginning to really start to make . . . you know, make a little bit of effort
> . . . you were putting extra effort into working. And I feel at the moment
> . . . I know there's lots of work on but you've had lots of help, haven't you?
> It's your turn now isn't it? To turn round and say "I can do this" and this
> is one of the things that you're good at. You're as good as anyone else.
> In fact you're better than a lot but you're slowing up again, so you're
> falling behind. And that's sad. Now come on, let's try and do it. . . .
> *(to whole class)*: Now there's lots of smiles on the faces of people who
> are well behind, I'm afraid. And it seems to be getting to be a bit of a

habit. We've got a lot of work to do this term and I'm afraid some of us ... if you're not up to mark and you're not actually working as hard as you should, you'll end up either being moved so that people who can make more use of the pace of this set ... into a slower set. I hate to say this, but it's true. People in Mrs. Camden's class *(a lower set)* are working a lot harder than you are. Or you'll end up staying in at playtimes and catching up with everybody else. We've got too much learning to do this term. We've got a lot of exams after Easter ... a lot of exams. And you need to know it and if you're not going to put your minds to it then why should Miss Black or I bother, quite frankly. When there are children in Mrs. Camden's class who are working much harder than some of you who deserve their place in this set. I felt it yesterday and I'm being honest with you today ... I felt it yesterday, I felt a lot of you are not working as hard as you should be. So just think on about that. Do you think you deserve your place in this set? I think some of you are gonna have to prove it.

(Black, 2002b)

In this speech, Mrs. Williams vocalizes not only an emphasis on speed as the basis for work in the set, but also the assessment background which is such a powerful driver for her. Here she again attempts to galvanize the class by naming particular children who have gained merit awards for their performance:

T: *(standing in front of class)*: I gave two stars to Toby cos he worked exceptionally well last week ... very good indeed and Jeremy and Carl and Chris. Not because he quite got as far as the others but because what he'd done was good and you're answering more in class. He often answers in class which is good ... he'll often have a go. Anybody else get two? Two's? *(no response)* Yeh about four of you. Well done all of you anyway.
Nelson *(to himself)*: I got one.

(Black, 2002b)

Although we see here a genuine attempt by Mrs. Williams to emphasize effort over speed and ability, the overall message of the classroom seems to be one of competition and coverage, and praise for those who work faster and get the answers right—ultimately, Toby has earned his second star *in spite of* his slowness. We can speculate further that this speech has done little to help Nelson, whose comment to himself suggests discouragement in the lack of appreciation for his one star. This emphasis on speed, as Horn (2007) argues, is difficult to overturn: in her research in schools which are explicitly committed to mathematics education reform (which is not at all the case in Mrs. Williams' class), teachers struggled to move away from constructing learners as fast, slow or lazy (with "fast" equaling "good") towards a reconstrual of mathematics learning in which "fast" can in fact equal *not* good: "speed may also be a liability, as fast kids 'often [are] not stopping to think about what they're doing.' . . . Their focus on task completion may be adaptive to schooling, but not necessarily to complex thinking" (p. 56). Indeed, the conflation of speed with doing well, even at the expense of understanding, is demonstrated by this exchange between Nelson and Carl:

Nelson: Daniel, aren't I fast at finishing my Richmond test?
Daniel: What?
Nelson: Aren't I fast at finishing my Richmond test? I just went . . . I was like zoom through it, it's well easy. Well not all of it was easy cos I still got some wrong.
Carl: Did you?
Nelson: I guessed some. Did you guess any?
Carl: One or two. How many did you guess?
Nelson: Dunno.
Carl: I guess a few, something like four or five. Ten?
Nelson: Ten to fifteen, around that. And I got some guessed right so. When I finished the whole thing . . .
Carl: That's the whole point in having it again.
Nelson: I'd guessed about ten of them.
Carl: Did you? They're easy.
Nelson: I know, I just had to finish before Jeremy. And Jeremy didn't even guess, he still gets it right.

(Black, 2002a, pp. 234–5)

Unsurprisingly, speed is central to the children's positioning of themselves and of each other, and their talk is dominated by constant requests for updates on what question number others are doing, accompanied by defensive and/or aggressive comments on the significance of this for how good at mathematics they are:

Nelson: That's the answer, seventy-five. Are you stuck on number twenty-three or twenty-four? Twenty-four: "Janet buys a place for . . ."
Chris: What are you on?
Nelson: Erm twenty-four. Are you on B1?
Chris: No.
Nelson: Twenty-four?
Chris: Yeh.
Nelson: Believe it or not but I'm in front of you.
Chris: What?
Nelson: Believe it or not but . . .

Chris: Only cos I keep rubbing out everything
(later)
Nelson: How does Erica get that far?
Chris: She's only on three past us. She was about . . . about ten past us yesterday.
Nelson: She's hardly ever past us. Is she?
Chris: She must just be good at this.

(Black, 2002b)

The children are inevitably competitive. Here Nelson, Erica, and Carl compete while once again positioning themselves with respect to Jeremy:

Nelson: What are four fours? Carl what are four fours? Carl what are three fours?
Carl: Why?
Nelson: Cos I need to add 'em. *(pause)* So you don't know. Carl doesn't know what three fours are.
Carl: Yes I do.
Nelson: Carl doesn't know what three fours are. Four, eight . . .
Carl: Why did you ask me then?
Erica: Nelson doesn't, that's why he asked.
. . .
Carl: Who's on number six?
Erica: I'm on number five.
Nelson: I'm on three. You know Jeremy? Jeremy's on this.
Erica: On that? I'm on number B.
Nelson: Carl, Jeremy's on this. Jeremy's on this page. He was . . . you know when he went on to this when I went, he was there. You were on B1 . . . you started on B1. Jeremy went to Miss, he asked Miss for help on number eight I think or some . . .
Carl: I need a ruler, have you got one?
Nelson: I'll get you one if you tell me number three.

(Black, 2002b)

The following exchange between Jane and Sian shows similar levels of competition and comparison, embedded in claims about how easy they are finding the task:

Jane: Easy, god, this is infants' stuff.
Sian: Boring.
. . .
Jane: Have you carried on the tag or have you just done it from there? Sian, have you carried on the tag or have you done it from there?
Sian: Carried on . . .
Jane: I'm gonna carry on too.

Sian: Just because I've done it.

. . .

Sian: What number are you on?

Jane: Six.

Sian: I'm on twelve.

Jane: Cos you started before me.

Sian: Yeh but we both started the same time actually.

Jane: I'm on number seven now, I've finished.

. . .

Sian: It's not too bad, maths. This is dead easy.

Jane: Maybe they think it's hard. *(Refers to another pupil in the class)* [She] thought it was hard because she's a bit tired. Though it wasn't her fault she didn't understand, do you think?

(Black, 2002b)

The complexity of the dynamics of positioning is illustrated by various other events in the classroom. Not all pupils in Black's data set have consistent positions in the classroom; some individuals can be seen to gradually shift position in ways which appear to be linked to ongoing talk patterns, mostly in the shape of particular teacher feedback and/or critical events as they follow a trajectory which is influenced both by moments within the classroom and longer-term personal history or characteristics. The following contrasting exchanges from this data set demonstrate such a shift in the case of Janet; both are discussed further in Solomon and Black:

Janet extract 1: What fraction of a number of egg-cups contain an egg?

Janet: Do you only count the eggs that are in the egg cups?

T: What?

Janet: Do you only count . . .

T: Ooh no because altogether you've got six. How many have you got?

T and Janet: Five.

T: So that's the whole . . . is five. So each one of those is a fifth. And how many out of five are in the egg cups?

Janet: Three.

T: Three out of the five.

Janet: So that's three . . .

T: Three fifths.

(Solomon & Black, 2008, np)

As in the extract above, in which Phillip's misunderstanding is treated as temporary and the conversation rapidly repaired, Janet's initial lack of clarity about the task is responded to with a recasting (although not total clarification) of the situation. The teacher goes on to scaffold Janet in a duet style in which

Janet maintains her agency in the dialogue, offering a final "so that's three . . ." check on her own understanding. This maintenance of an active part in the conversation is however missing from the following extract, which was recorded at a later date:

Janet extract 2: What number of eggs represent a given fraction?
T: Eh number six, what is four sixths of four . . . of . . . what's four sixths of six eggs? So what's the first thing to do . . . we do Janet? To find out four sixths of six eggs. Got to find out . . .
Janet: Four.
T: Not four, the bottom number remember is telling you how many parts you're dividing it in to, so which number do you share it by? (*Other pupils have hands up*)
P2: Six.
T: Six yeh it's the six underneath, it's the bottom part of the fraction, isn't it, that's telling you how many parts you're sharing, whatever the number is, whether it's a hundred or twenty, a two, a six or eight or one thousand five hundred and sixty-five, it's still being shared between that . . . into that number of equal groups is it not? (*teacher is animated with hands here*)
Janet: Yeh. (*quietly*)
T: Yeh so you're sharing it by six, how many sixes are in six? (*pause*) one. So you'd have one, yeh, so what would four sixths be then?
Janet: Four.
T: Four . . . yeh . . . is that what you were telling me before? Was it the answer you were giving me? I'm sorry (*sympathetically*).
(Solomon & Black, 2008, np)

Janet's initial "four" is misinterpreted by the teacher as irrelevant, and she does not seek to repair the situation, only realizing her error some time later. Here we see Janet reduced to monosyllables in a controlling interchange which is much the same as Hasan's above. It is noteworthy (see Black, 2002a and Solomon & Black, 2008) that subsequent interactions involving Janet were in fact predominantly non-productive ones, and that her positioning in the class had undergone a significant shift.

The reverse situation occurs with James, however. Again, the following two extracts, taken from Black (2005, pp. 14–16) are separated in time:

James extract 1: How many complete sets of nine skittles can be made from a given number?
T: This time you concern yourself with the nine times table. "How many complete sets of nine skittles can be made from these?" So we're sharing by nine. What did we do yesterday when we wanted to share by nine, James?

(*pause*) Any number, how can you do it? You can't halve it, can you? (*pause*) To make it easier for yourself. What have we got to do with those big numbers, like a hundred and fifty-eight? (*pause*) What sort of sum are you going to have to do whether you like it or not?

James: Erm half it and then put a remainder.

T: Ooh are you halving it? If you're halving it what number are you dividing by? (*pause*) If you half something how many groups?

James: Two (*quietly*).

T: You're dividing by two. We're sharing by nine.

James' interaction has the hallmarks of a Group B pupil here, as Mrs. Williams fires questions in quick succession to open and then again in response to his (incorrect) contribution, which appears in fact to be prompted by something the teacher has said in her opening remarks ("You can't halve it can you?"). Her stream of pseudo-questions regarding halving suggests that, rather than perceiving this as a situation in need of repair, she takes it as a signal that James is in need of remedial teaching, with the result that he is quickly reduced to a single-word and quiet answer. Her final response asserts her status as primary knower and his as secondary knower, to use Martin's (1992) and Berry's (1981) analyses. James did not maintain this general position, however. Black reports that his interactions increase over time and become more active, as in the next extract:

James extract 2: Introducing the concept of factors.

T: We're going to be looking at some new different types of words, for example, does anybody know what a factor is? If I said, give me a factor of fifteen, can anyone give me one? (*pause*) . . .

James: I . . . I kind of understand.

T: Go on then, you tell me what a factor of fifteen is.

James: Is it . . . is it a number that'll add up . . . erm times or add up to . . .

T: (*interrupts*) Times . . .

James: Up to fifty?

T: Up to anything. Up to any number. So what's a factor . . . give me a factor of fifteen then. (*Sean puts hand up*) So what number will divide equally into fifteen? (*she nods at Sean*)

Sean: Three.

T: Three. That's a factor of fifteen.

James (*calls out*): And five.

T: Think of another . . . five. Think of another one? (*nods to Chris*)

Chris: One.

T: And another one? (*pause*) (*Sean has hand up*)

Sean: Fifteen.

T: Fifteen. Those are the factors of fifteen.

Here we see James acting more like Phillip and the teacher responding to him in a very different way to extract 1. First, James positions himself as an active participant by volunteering, and this is taken up by the teacher. Second, we see here a very different reaction to James' error and hesitancy in his "is it a number that'll add up . . . erm times or add up to" from what we might have anticipated in extract 1. Rather than responding didactically, with evaluation or with pseudo-questions, the teacher offers the correct answer ("Times") with no more ado—she is keeping the dialogue going.

The example of Janet has considerable similarity with a scenario described by Gee (2001), who suggests four views of identity: Nature-identity, Institution-identity, Discourse-identity, and Affinity-identity. The first refers to those identities that are based on a state of being that is determined by nature, not society (his example is "being a twin"). Institution-identity is based on authorization by institutions and refers to position within an institution, such as being a teacher. A Discourse-identity draws on discursive processes and the dependence of identity on interactions with other people. Finally, an Affinity-identity draws on experience within a particular group. Gee (pp. 116–19) tells the story of an African American girl who "bids" (p. 118) through her keen participation in classroom talk to get herself recognized as a "proactive and enthusiastic learner" and to *achieve* a corresponding Discourse-identity, but has this rejected by the teacher who by her response "invites" her to take an *ascribed* Discourse-identity of "the sort of learner who needs to be managed by the teacher's instructions" (p. 118). Gee's analysis is that she accepts this invitation and subsequently acts out this identity as an *achieved* Discourse-identity. It is interesting to consider how it is that the teacher has the power to do this, and we might conclude that her power arises primarily from the institution and the Institution-identity that she invokes.

Indeed, the force of an imposed Institution-identity is demonstrated in the same group of children in the case of another African American child, this time a boy who, despite reading above grade level, is positioned as an "at risk" learner who has the institutionally recognized features (black, poor, black vernacular speaker) which will trigger a reform package. Again, he is "invited" to take on the Institution-identity which has been prepared for him. As Gee says, the children can resist these ascribed identities but the question arises as to how they can do that—perhaps by taking on a particular Affinity-identity, but only within a constrained set of possibilities and the resources that they afford. This analysis illustrates how Gee's four identity perspectives intersect to position and enable/induce/produce self-positioning on the part of the individual in a Discourse ("any combination [of speech, action, physical expression, dress, feeling, beliefs and values and tool use] that can get one recognized as a certain 'kind of person'" (p. 110). Our particular histories and experiences and our narration of these combine so that:

Each person has a unique trajectory through "Discourse space." That is, he or she has, through time, in a certain order, had specific experiences within specific Discourses (i.e., been recognized, at a time and place, one way and not another), some recurring and others not. This trajectory and the person's own narrativization (Mishler, 2000) of it are what constitute his or her (never fully formed or always potentially changing) "core identity." The Discourses are social and historical, but the person's trajectory and narrativization are individual (though an individuality that is fully socially formed and informed). (p. 111)

Returning to the case of Janet, then, we can frame this as another example of failed bidding which results in an "invitation" from the teacher to take on the Discourse-identity of needing to be managed which she does indeed seem to accept—the second extract appears to represent a crucial episode after which Janet's behavior in the lesson moves on a steady trajectory towards non-participation. Why exactly Janet is re-positioned as she is we can only speculate, although as Black notes, she is Chinese and new to the school, and she is female. James, on the other hand, bids for the role of genuine interlocutor, and this is accepted—Black considers this to be a result of a display of cultural capital during the course of the observation period in which James told the teacher how his father had helped him at home with his fractions homework. Black suggests that this is a critical incident:

James unconsciously uses a positioning strategy by stating publicly that his father has assisted with his homework and has discussed the desired rule on calculating fractions. In doing so, he displays to the teacher the possession of "legitimate" knowledge which he has at his disposal ... James raises his status in the teacher's eyes by suggesting he has the resources for participating in the practices of school mathematics ... [He] exercises his sense of agency to re-position himself as someone who is able to make a more active contribution to the local practice of learning school mathematics. (Black, 2005, pp. 17–18)

Viewing the primary school data within this analytic framework, it is possible to see that there are cultural influences and institutional structures which underpin the range of positions available to learners in the mathematics classroom. These positions are maintained by and for individual pupils through their on-going participation in particular kinds of classroom practices in which both they and their teacher draw on a range of cultural resources and discourses which contribute to the emergent patterns that we see. I consider the role of discourse in these emergent patterns in more detail in the next section.

Looking at Emergent Patterns—Discourses of Gender and Ability

One of the major influences on the emergent classroom culture is that of what pupils and teachers bring with them in terms of cultural "baggage" concerning gender and mathematics—their self-positionings, and their associated beliefs. They also bring with them values and assumptions concerning ability and the nature of learning. So, in addition to the long-term culturally-embedded and typical patterns of classroom discourse that I have considered in the previous section, there are other long-term influences on what happens within any individual classroom which have more to do with identity as we normally think of it. To put this another way, the dynamic of the classroom is one in which individuals identify with the discourses of gender and ability which are made available to them through the socially accepted and expected patterns of pedagogic interaction.

Gender and Pedagogic Discourse

A wide range of research has noted and tried to explain (and counter) gender differences in the quantity and quality of classroom interactions that pupils experience, pointing to the gendered nature of language and language use, for example in terms of women's deference to men in conversation (Spender & Sarah, 1980), and boys' readiness to use semi-technical registers and their control of argument genres (Corson, 1992). Focus on the nature of classroom discourse has suggested a cluster of related differences: Swann and Graddol (1988) reported on boys' dominance in primary classrooms in terms of amount of talk, turns taken, and exchanges with the teacher; Bousted (1989) remarked on 15-year-old boys' willingness to contribute and to interrupt; and She (2000) noted the passivity of girls compared with boys in whole-class biology lesson discussions. The common theme, then, is that boys take more than their "fair share" of the floor (Howe, 1997), and researchers have suggested that their talk shows typical characteristics—contributing individually rather than collaboratively and acting as dominant speakers (Coates, 1997)—as does girls' talk—tending to consensus and quietness (French & French, 1984) and reluctance to speak in public (Baxter, 2002).

On the basis of the discussion of classroom practices and pedagogic discourse above, we need to note that perhaps the teacher attention received by such boys may not be conducive to learning (i.e. it may be classroom-management focused rather than learning focused, an important distinction that French and French fail to make). However, research which examines this distinction suggests that in fact boys' greater participation can be seen across the whole of classroom talk, including learning episodes as well as classroom management (AAUW, 1995; Dart & Clarke, 1988). More specifically, in much the same way as Black's research shows, Good, Sikes, and Brophy (1973) and Swann and Graddol (1988) report that boys are more likely to engage in long turns of explanatory talk, compared with shorter factual statements from girls, and they receive more praise. Sadker and Sadker's (1985, 1986) research found similarly that boys were listened to by teachers when they called out, whereas girls were referred to the classroom convention of raising a hand in order to speak. Eccles (1989) also found that a handful of boys in each mathematics class in her study received particular attention in comparison to other students.

The main issue, however, lies in how these differences are explained. An initial reaction is to lay the blame on teachers, given that their own domination of the classroom means that pupils are in fact responding to what teachers do. There are indeed reports of teachers' overt sexism (Delamont, 1990) but also of rather more subtle effects which teachers may well be unaware of. Swann and Graddol (1988) report that boys are more likely to be invited to answer questions, and, similarly, Stanworth (1983) argued that boys are the focus of teacher–pupil interactions while girls are marginalized. However, Swann and Graddol (1988) also note that such invitations may well be non-verbal (and possibly therefore less conscious), that boys are more likely to be first to put their hands up, thus triggering teacher choice, and also more likely to volunteer contributions in informal situations where speakers are not being selected, while girls for their part *expect* a lower participation rate. Indeed, Corson (1992) suggests that these observed differences are a result of gender socialization and different cultural interests which place boys at an advantage when it comes to talking in the classroom.

In the same vein, She's (2000) investigation of secondary school biology classes concludes that male dominance is perpetuated through an interaction between boys' "natural" behavior of calling out answers and teachers' beliefs about gender differences in learning styles which mean that they do not attempt to control such behavior and instead direct more questions and related feedback towards boys. Bousted (1989) also argued that classrooms are sites of social reproduction, specifically the public/private distinction in the lives of men and women, and in this respect it is noteworthy that girls are likely to seek individual teacher attention (Dart & Clarke, 1988). More mundanely, it can be argued that boys are simply more visible due to their generally higher level of physical activity, thus inviting more attention (Howe, 1997). However, as research and theory have progressed it has been recognized that socialization and difference arguments over-simplify the situation, not least because of their assumption of the homogeneity of behavior within one gender and also the assumed passivity of girls. Thus Stanley (1993) argued that quiet girls were silently resisting rather than passive victims, while Hammersley (1990) pointed to the over-simplification of boy/girl comparisons which ignore individual differences, perceived ability being one of them, as indicated by Good et al. (1973). Thus a close look at French and French's (1984) data reveals that although boys take more turns than girls, these are in fact simply a measure of the activity of a sub-group of boys.

Assessing these findings and analyses on a theoretical level, it is possible to discern a number of approaches. One such is an assumption of essential differences between boys and girls in preferred language and learning styles—an example is provided by Jenkins and Cheshire's (1990) examination of girls' and boys' performance in the oral examinations which were a compulsory component of the English Language GCSE at the time. As Baxter (2002) points out, this approach makes assumptions about basic differences: "female talk is characterized as a co-operative, facilitative, supportive style of engagement more suited to private and informal contexts, whereas male talk is likely to be more competitive, adversarial and authoritative, and more geared towards formal or public settings" (p. 82). The implications of a strong stance on difference are that girls learn differently and should be taught differently—I will explore this idea further with respect to mathematics education in Chapter 7. An alternative but related stance exemplified in Spender's (1980) work is that which notes the implications of the public/private split, and argues that boys have an advantage in terms of the valorization of public speaking because of their preferred style: if women are to compete they must adopt "unnatural" masculine styles. Other research such as Swann and Graddol's (1988) and Delamont's (1990) invokes more of a sex-role approach in terms of its assumption that girls conform to their socially ascribed roles and thus have little expectation of public speaking.

However, these approaches tend to treat both boys and girls as passive: writing at a later date, Swann (1994) advocates contesting inequality and fostering girls' resistance. The complexity of such a project is explored by Davies (1990) in her examination of the need for discursive, personal and social resources in any assertion of agency and choice within the range of available discursive positions in the classroom. As Baxter (2002) suggests, while "effective public talk" is a generally available discourse, its accessibility is not the same for both genders, and is dependent on the different ways in which speakers are positioned, and position themselves, within the classroom. Girls and boys participate differently in classroom discussions because they are taking up, negotiating and maintaining those positions which are open to them, drawing on particular classroom discourses to do so.

In drawing attention to these issues of agency and individuality in pedagogic discourse, it becomes clear that the patterns that we can see in classrooms constitute, and are constituted by, the ongoing cultural expectations and available discourses—or Discourses, in Gee's sense. As Creese, Leonard, Daniels, and Hey (2004) suggest, "classrooms allow children to 'shift positions' (or not) by virtue of a school's specific values, pedagogies and discourses" (p. 192), and it is in this sense that gender identification and pedagogic discourse intersect. As in the case of Gee's African American girl or Black's example of Janet, some of the available discourses are more accessible and "culturally appropriate" than others. Thus boys and girls take up particular gender positions and behaviors within this context. In one of Creese et al.'s middle-class classrooms, for instance, a dominant discourse of individual learning, "a discourse of difference and uniqueness and self-confidence" (p. 197), is readily taken up by the academic middle-class boys but also by the parallel group of girls who were able to "extend" their range of competencies beyond the collaborative practices which are identified as common to girls, to include the confident individualism more usually associated with boys. Exceptions are, predictably perhaps, the working-class girls, who remain in a different role, collaborating closely and keeping quiet. Importantly, this collaboration is seen by the teacher (and the middle-class boys) as not working at the height of their potential, merely producing average work as a result.

In contrast, Creese et al. describe a school which has a dominant discourse of collaboration and respect for diversity and difference, with boys' talk characterized by clear turn-taking and tolerance of multiple viewpoints, reflecting the pedagogic values of the teacher and the school. Unlike the children in the other school, these pupils were skilled at collaborative work, and mixed ability and gender groups are described as respectful and confident. Nevertheless, there are important gender differences in the available discourses within this context which center on the issue of "helping." Thus Creese et al. report that the boys describe the girls as "helpful" (p. 202) and that indeed this is a feature of the teacher's pedagogic aims. They argue that the teacher has been successful in encouraging boys to use peers as a resource (as opposed to working on their own) but there is something of an imbalance between the genders, it seems, as this exchange suggests:

Tabitha: I find that sometimes, only sometimes because most of the time they [boys] will work it out for themselves, but occasionally if you're sitting next to a boy and you've been put next to him, they'll say "Oh I don't know this question. Can you do it for me?"

Wyn: Yeah, it's like . . .

Grace: That's not true.

Tabitha: Only sometimes.

Wyn: Only sometimes.

Grace: They will ask you to help them, I'll ask them to help me, but they don't ask you to write it down on paper.

Tabitha: Not write it down, but give them the answer.

Wyn: Yeah, they do, they do. If they can't be bothered they will ask you.

Tabitha: I'm not saying that they are saying to do it exactly, but they, if they are in . . . sometimes, I mean this happens very rarely, but they will sometimes say "Can you just tell me the answer?" Say they're behind on their work or something.

Grace: They don't do that.

(Girls' friendship interview, Cityscape) (p. 202)

Although it may well be the case that the teacher has taken up "an equality position with which she attempts to shift identity positions and challenge normative male gender behaviour" (p. 202), it is less than clear whether she has challenged normative female behavior. The girls' position is ambiguous here and, as Creese et al. note, the interpretation of this exchange as challenge to social expectation or acceptance of an equality discourse is difficult, but the net result seems to be that these girls identify in a familiar female role.

While Creese et al. concentrate on the discourse of group work, their focus on friendship groups alerts us to another source of classroom-based discursive practice. Baxter's (2002) study provides a powerful example of the role of peer-based discourses in the construction of particular classroom identities. She notes that particular (i.e. more confident) boys take up more powerful positions within the classroom, utilizing the available discursive practices to construct an "effective speaker" identity. They do this by means of their self-positioning as "jokers" who perpetrate peer- and teacher-approved humor and their enlisting of peers to support this role. The boys' collective activity is such that they are able to control the discursive space and effectively silence disagreement in classroom discussion, particularly from girls.

Baxter also reports on a classroom discourse of gender differentiation in which the girls are constructed and construct themselves as loyal to each other such that taking on a leadership role is seen as "out of order" whereas it is quite acceptable for male students to be dominant speakers in public settings— "female leadership is still a highly contestable construct within a patriarchal society ... females have themselves constructed taboos against female competitiveness for positions of power, in a world where male leadership is still regarded as a cultural norm" (p. 93). This discourse dovetails with that of humor, which, while *not* being the province of girls, "tends to act primarily as a *conformist* dynamic in reinforcing male heterosexual masculinities in relation to 'weaker' or more subordinate 'others,' such as girls, 'boffins' [stereotypical 'experts'], or boys who are seen as 'gay'" (p. 90).

Analyses such as those presented by Creese et al. and Baxter suggest an explanation for the apparent anomalies in studies of gender differences which show, as French and French (1984) do, for instance, that not all boys dominate, and, as Stanley argues, that quiet girls are not necessarily passive:

Speakers are produced within different discursive conditions according to categories such as age, race, class, gender and so on ... Ironically, in order for the more dominant and popular boys to become "jokers in the pack", the support group boys willingly appear to take up, or at least, are prepared not to resist, their own subordinate subject positions. In other words, it is not just the majority of girls who tend to defer to dominant boys in the classroom public arena, it is also a significant number of less popular or confident boys. ... Boys are not uniformly powerful in the classroom, nor girls uniformly powerless. (Baxter, 2002, p. 94)

These studies demonstrate, then, that the relationship between classroom practice and successful learning is not straightforward. Despite an invitation by the teacher to enter into the kind of dialogue that Phillip enjoys, for instance, learners such as Hasan, or even Janet later on in her classroom career, position themselves in such a way as to take a very different role in the exchange from Phillip's. What is crucial is *what has gone on before* in terms of the interaction between classroom discourse, the implicit and explicit values of the institution, peer group positionings and gender identification processes. As Creese et al. note,

> Different pedagogies expose boys and girls to different learning discourses and therefore to a variety of gender identifications . . . gender is not determinate of who does well in each classroom. Rather, it is the discourses of pedagogy and the classroom cultures they create which allow particular groups of boys and girls to do well while others are less successful. (p. 194)

We need to add to this mix the peer group discourses which are also part of individual self-positioning. One such discourse focuses on the concept of "ability" and its embodiment in the practice of teaching groups of similarly performing students—variously called setting, streaming, banding and tracking—especially in mathematics classes.

Discourses of Ability

The discourses and practice of mathematics that I have reviewed in Chapter 1 include and endorse some powerful beliefs about ability and, relatedly, about the supposed inherent difficulty of mathematics and the personal qualities that are required to do it successfully. Furthermore, the ultimate manifestation of success is not merely getting right answers, but getting them quickly and with an apparent lack of effort which is associated with having a natural aptitude (although mental instability is perfectly allowable, as Mendick (2005a) points out in her analysis of choosing mathematics). Doing well in mathematics requires serious identity work which among other things is fundamentally gendered:

> "being good at maths" [is] a position few men and even fewer women can occupy comfortably . . . "real mathematics" is different from other subjects; it is certain and rational; "real mathematicians" are different from other people; they combine the flattering character of geniuses and heroes with the unflattering character of "nerds". These discourses are oppositional and gendered; they inscribe mathematics as masculine, and so it is more difficult for girls and women to feel talented at and comfortable with mathematics and so to choose it and to do well at it. (Mendick, 2005a, pp. 216–17)

The AAUW's *How schools shortchange girls* (1995), massing together a large number of studies, similarly reports on the difficulties for girls in envisaging themselves as successful in mathematics because it is perceived as "masculine." Whether or not girls pursue advanced study is correlated with levels of confidence rather than competence, and girls are more likely than boys to attribute failure to ability while boys are more likely to attribute success to ability (Fennema & Leder, 1990). The AAUW report also collates evidence that girls who express self-images of competitiveness are more likely to follow through to higher mathematics, as do those who reject traditional gender roles. Similar effects are captured by Steele's (1997) notion of "stereotype threat," which seeks to explain why women who identify as good at mathematics nevertheless remain uncomfortable with this aspect of their identity, running counter as it does to culturally-based assumptions.

In the institutional context, discourses of mathematics ability are lived out and gain major currency within a pervasive practice of grouping by performance ("setting" in the UK) which often begins in the later years of primary school and is highly visible in the secondary years. This practice is powerful and largely uncontested within UK education policy and practice spheres, despite the fact that researchers have reported for some time on its adverse effects (see, for example, S. J. Ball, 1981). For many, their perceptions and expectations of themselves, and their relationship with mathematics revolve around their set membership. In England and Wales, this is felt particularly keenly by GCSE students in terms of what tier of the GCSE examination they are entered for (Boaler & Wiliam, 2001; Boaler, Wiliam, & Brown, 2000). Set membership is salient not only during the school years, but remains an identifying feature of adult feelings about mathematics (Buxton, 1981; Swain, Newmarch, Baker, Holder, & Coben, 2004).

A range of research indicates that mathematics teaching sets are characterized by particular cultures and self-positionings which are heavily gendered and classed. For example, the "top set" in Bartholomew's (1999) research is distinctive in that a particular subset of (middle-class) boys is positioned as the teacher's equals, set apart from the other pupils by their confidence, speed and apparently effortless achievement of right answers—they are "budding mathematicians," labeled as such by their behavior and set membership rather than their actual performance in some cases. While other pupils work hard to "earn a place" in the top set, these "high fliers" do not need to justify theirs. Girls in top sets were likely to be positioned and position themselves as having "less right" to be there and to experience a high level of anxiety (see also Boaler, 1997b; Boaler et al., 2000). These differences between mathematics set cultures have considerable consequences for student identities. As Bartholomew (2000, p. 6) argues, "the culture of top set maths groups, and of mathematics more generally, makes it very much easier for some students to believe themselves to be good at the subject than for others."

The actual composition of teaching sets shows patterns which appear to be a reflection of the underlying discourses of ability concerning *who* is able at mathematics. At secondary school levels, research by Boaler (1997a) indicates that working-class pupils are more likely to be placed in lower sets, even after controlling for previous performance, a finding echoed by Wiliam and Bartholomew (2004) and Bartholomew (1999), who also notes that girls are more likely to be placed in lower sets. Gillborn and Mirza (2000) found a parallel effect in the allocation of ethnic minority pupils to lower sets, while US research shows that poor students and students of color are under-represented in more demanding classes (Ladson-Billings, 1995, 1997; Oakes, 1990). Boaler (1997a) observed that correlations between secondary school entry attainment and GCSE outcomes were weaker in a school which grouped by performance than in a school which did not—for example, students who achieved better GCSE results than their entry attainment scores would predict tended to be middle-class boys; the converse group of pupils achieving *lower* grades than expected were mainly working class. As a whole, this group showed a bias towards girls, although within the working-class majority the gender split was more evenly balanced.

These observations have considerable implications for these students' access to mathematics. As Boaler (2002, p. 132) points out, a situated perspective on learning underlines how "different pedagogies are not just vehicles for more or less knowledge, they shape the nature of the knowledge produced and define the identities students develop as mathematics learners through the practices in which they engage." Despite the pervasive belief that "ability" grouping is (a) necessary from a pedagogic point of view and (b) raises achievement (see Ireson & Hallam, 1999), there is little evidence for this view—the net result from meta-studies in both the UK (Ireson, Hallam, Hack, Clark, & Plewis, 2002) and the USA (Slavin, 1990) is that there is little effect, but that if there is any it is in the direction of adverse effects on the lower set students. Exploring the issue in depth, S. J. Ball (1981) provided evidence that sets differed in the type of teaching that pupils experienced and the kinds of relationships with teachers that they had—lower sets received less discussion and more boardwork, and a reduced curriculum. This remains true. Wiliam and Bartholomew's (2004) analysis of setting, teaching styles and examination performance in mathematics in six schools leads them to conclude that "the most pernicious effects of setting may not be necessary consequences of grouping students by ability, but appear when teachers use traditional, teacher-directed whole-class teaching" (p. 289). Although in all the schools, students in top sets made greater progress than those in lower sets, the discrepancy in progress was greatest in those schools which used whole-class teaching. In these schools, Standard Attainment Tests (SATs) performance in Year 9 (age 13–14) did not predict GCSE performance in Year 11 (age 15–16), whereas set membership did. In contrast, in schools using predominantly small group or individualized teaching, top sets did not perform significantly better at GCSE in comparison with pupils in lower sets when SATs attainment was controlled for.

Once in lower sets, pupils are likely to experience a "polarized curriculum" which limits exposure to mathematics and the GCSE grades they might attain (Boaler & Wiliam, 2001; Boaler et al., 2000). This polarization means more than a quantitatively restricted curriculum however: there are qualitative differences as well as Morgan's (2005) research in Chapter 1 and Dowling's (2001) work, which I will discuss in Chapter 7, shows. Bartholomew's (1999) research suggests that there is a strong likelihood that teachers will have lower expectations of pupils in lower sets, and that these expectations correlate with different kinds of teacher–pupil interactions. Faced with higher sets, teachers were more likely to focus on pupil learning and involvement with the subject. A complex version of this phenomenon is also reported in Horn's (2007) research, where teachers were concerned to work out how to deal with what they perceived as a mismatch between the move towards heterogeneous classes and equal access to a high-level mathematics curriculum, and student performance:

> ... teachers often see their students' prior achievements as incommensurate with a rigorous mathematical curriculum—and, in many cases, students themselves concur, only adding to the teachers' challenge. ... Teachers often view cognitively demanding modes of instruction as inappropriate for low-achieving students ... (pp. 42–3)

The problem, Horn argues, lies in two important assumptions about mathematics: that it is a body of knowledge which must be taught sequentially and with full coverage of certain topics before others (with implications for speed of learning), and that collaborative learning is only workable with high-performing students. Together, these assumptions make it difficult for even reform-committed teachers to develop a pedagogy based in equity, while "slow learners" for their part continue to play their designated role as low-status members of the class.

The reciprocal nature of the relationship between setting, self-positioning and corresponding access to mathematical knowledge is particularly well demonstrated by Zevenbergen (2005). She argues that lower performing students' awareness of the restrictions on them in terms of curriculum and pedagogy lead them to develop a predisposition towards mathematics as negative and to behave in ways which contribute further to their reduced participation in a reciprocal relationship between the "ability-grouping" practice and their own experiences:

> In the case of the lower-stream classrooms, the students' reactions to what they see as inferior or poor experiences create a *habitus* whereby they

behave in particular ways in their classrooms. They are creating a *mutual construction* of the classroom ethos: their behaviours can be seen as contributing towards a particular structuring practice—one which will contribute to their performance in mathematics. (p. 616)

In contrast,

> The higher streamed students made it clear that they have a greater sense of belonging to their classrooms than their peers in the lower streams. They are more positive about mathematics, and have a greater sense of being able to achieve in the subject. . . . their experiences had enabled them to construct a mathematical *habitus* whereby they perceived themselves as well positioned in the study of mathematics . . . They have come to see themselves as clever and worthy of their positive experiences. (p. 617)

This effect becomes particularly strong at the post-compulsory level: Mendick (2006) reports that in the further mathematics classes she observed, the teacher emphasized that the further mathematics group is "different" and should be able to move more quickly through the curriculum than the regular mathematics group. This gives time for more theoretical work which is not in fact examined, but the important message (for teacher and students) is that the students should know the principles underlying what they are learning, and that this matches their ability, assumed to be greater than the other group. This distinction is important and leads not only to curriculum differences but to pedagogic difference: thus Mendick (p. 129) describes how one teacher makes assumptions regarding the further mathematics students which prevent him from adjusting his practice with the other group:

> His understanding of further maths students as different, and importantly as more able, enables him to teach them less hierarchically and prescriptively but also prevents him from considering shifting the ways that he teaches other people maths.

It is in response to observations such as this that critics of traditional mathematics teaching emphasize the need for change on two related fronts: in terms of content, to include problem-solving, conjecture, experiment, argument and justification (D. L. Ball, 1993; Bauersfeld, 1995; Cobb, Yackel, & Wood, 1993; Lampert, 1990; NCTM, 2000; Steinbring, 2005; Williams & Baxter, 1996), and in terms of relationships, to involve the classroom community in supporting a participatory pedagogy which encourages exploration, negotiation and ownership of knowledge, all of which involve an identity shift for many learners (see for example Boaler, 1997a, 2000, 2002, 2008; Boaler & Staples (2008); Burton, 1999b; Fennema & Romberg, 1999). This characterization of mathematics brings with it an emphasis on the role of classroom interaction in learning and the teacher's role in it, underlined by the NCTM (1991) *Professional standards for teaching mathematics* as "eliciting and engaging children's thinking; listening carefully; monitoring classroom conversations and deciding when to step in and when to step aside" (Nathan & Knuth, 2003, p. 176). Supporting this kind of learning is then a question of fostering discussion between pupils, enabling exploration, and assisting learners in formulating their ideas in a mathematical way, an approach now advocated in the UK by agencies such as the Standards Unit which aims "to encourage a more active approach towards learning Mathematics and to develop more connected and challenging teaching methods" (DfES, 2005, p. 5). The Qualifications and Curriculum Authority (QCA) acknowledges similarly that "Mathematics is a creative discipline. It can stimulate moments of pleasure and wonder when a pupil solves a problem for the first time, discovers a more elegant solution to that problem, or suddenly sees hidden connections" (QCA, 2006).

But while much U.S. and European mathematics education reform, seeks, essentially, to "promote meaningful learning" (Nathan & Knuth, 2003, p. 175), this is not easily attainable. As a number of researchers point out, creating the classroom environment in which this can happen is more easily said than done. There are various issues to be negotiated in the development of good practice defined in this way, including teacher beliefs, prior experience and institutional practices (in the US, Nathan & Knuth, 2003, and in the UK, Ireson & Hallam, 2001), the legacy of a very different teacher training (Ross, 1998), the absence of specific guidance in reform pedagogy (D. L. Ball, 1993) and indeed the development of teachers' own identities (van Zoest & Bohl, 2005; van Zoest, Ziebarth, & Breyfogle, 2002). Students' own identities and self-positionings also contribute to the dynamic, and their acceptance of, or ability to engage in, different practices is not assured (Cobb & Hodge, 2002; Lubienski, 2007). The importance of identity is that, as Wenger's work indicates, it mediates the individual's relationship with the practice—in this case the practice of mathematics. What lies at the heart of that practice, as I have shown in Chapter 1, are its ways of making meaning, and these are inextricably linked with the identities of knowing as authority, expert, novice, and "legitimate peripheral participant" which are negotiated within classroom cultures. In the next chapter I explore how secondary school pupils describe their mathematics classes and their relationships with their teachers—like many before them they describe differences in experiences between and within sets which are enmeshed with perceptions of themselves as mathematics learners, and of the nature of mathematics itself.

3

Mathematical Moments
and Mathematical Lives
Doing Mathematics at Northdown School

In the previous two chapters, I considered how theory and research enables us to capture how particular experiences contribute to mathematical identities, and how learners position themselves, and are positioned, within the available discourses occurring in the context of classroom, institutional and peer cultures. In this chapter, I explore the development of individual mathematics identity trajectories, focusing on an analysis of secondary school students' narrativization of learning and doing mathematics. The analysis is organized by set (i.e. "ability" group) membership and focuses on the students' positioning and self-positionings with respect to the range of beliefs about ability, gender differentiation and the nature of mathematics as they occurred within these groups. The stories they tell thus provide a means of mapping their shifting mathematics identities as they draw on these various discourses to make sense of their experience. The Year 7 analysis acts as a "mixed abilities baseline" against which the Year 9 and 10 groups can be compared.

Mixed Ability Classes: The Year 7 Stories

Year 7 enter Northdown School accompanied by their recent experience of assessment in Key Stage 2 SATs at the end of Year 6, and their resultant scores in English, Mathematics, and Science. Once at Northdown, they are taught in heterogeneous groups, but they are aware that they will be grouped at the end of the year on the basis of their performance, moving into new teaching sets in Year 8. Their accounts were characterized by a number of strong claims about ability and effort, but also a certain amount of conflict: although they were clearly determined to rehearse an incremental view of ability and the related claim that effort reaps reward, they found this outlook hard to sustain in the context of their actual experience of mathematics learning. A related issue concerns their perceptions of the nature of mathematics and how it is learned: some described themselves as having little control over their learning, although others expressed more participative identities and personal epistemologies which suggested a corresponding level of ownership. Lying behind these developments is an indication of emergent differences in their experience of learning mathematics.

Competing Discourses of Effort and Ability

When the pupils in Year 7 were asked to consider what distinguished those who were successful in mathematics classes from those who were less so, they tended to point to an explicitly incremental view of ability. For example, Carol thought that those pupils who have been discouraged by failure in the past need to regain their motivation and belief in the potential benefits of effort:

> [They have] been put down because they've not been very good at it so they can't be bothered any more.... [They need to change] their attitude probably because they need to think "I can be as good as everybody else if I want to be".

However, the same pupils tended not to draw on the idea of effort in their more general talk about learning mathematics or when they were quizzed about how to help the less successful students. Having initially stated a belief in effort, Sylvia went on to contradict herself, saying that individuals differed, ultimately, in fundamental and unchangeable ways which did not respond to increased effort because they were constrained by natural ability:

> ... everybody is just like better at something different and understands different things ... you'd have to get them to listen more and try and take part more in group activities and concentrate a bit better [but] ... there's a lot of people can, do pay loads of attention and still can't do it ...

The tension between the publicly approved discourse of effort in the school and the mathematics department and the implicit or even explicit discourse of natural ability which permeated their actual experience of mathematics classes and tests, meant that such contradictions within their accounts were common. Within the context of the dominance of a sequential view of mathematics (Burton, 1999b) and its relationship with the discourse of "fast" versus "slow" which Horn's (2007) research reveals, this is perhaps unsurprising. Reminiscent of Reay and Wiliam's (1999) study and indeed of Mr. Wilson's emphasis on examinations in Chapter 1, most of the Year 7 pupils reported on both pupil and teacher stress in Key Stage 2 Standard Assessment Tasks (to the extent that in Nicola's school the teacher's distress had became a playground joke). The mathematics department was also, it turned out, under considerable pressure to "raise standards" (i.e. raise SATs and GCSE scores) due to low expectations in the past, and, together with the upcoming setting decisions for the following year, it seems likely that the students were aware of a certain amount of pressure to perform in tests, making it difficult to sustain an identity of being good at mathematics which incorporated effort and understanding rather than "natural ability," memory and, of course, speed. They were thus positioned by the traditional discourses of mathematics within a more narrow range of possibilities than the school's public ethos suggested.

This is borne out in practical terms when I asked them what they thought about setting—how was it going to be next year, and what would they prefer? Most had experienced setting at primary school as in Mrs. Williams' class, and all said that they preferred it. In fact, they were already experiencing it to some degree: the practice in the Year 7 classes was to group informally by performance within the class itself, and there was some evidence that these groupings were associated with different teaching and learning experiences which constrained the students' subject positioning with respect to mathematics. For instance, Carol and Nicola both tell me that they are members of "the group of eight at the back of the class," a well-defined group who are given different work to do because they work more quickly than the others. This group are positioned differently with respect to the usual teacher–pupil relationship norms: they are allowed to ignore him as he teaches the rest of the class as Nicola explains: ". . . there's about eight people who normally sit at the back of the class and they do work normally whilst he's explaining." They are also given tasks which indicate a different role within the classroom community, and which are explicitly linked to performance: Jake reports that "I sometimes get called off my regular class work because I'm one of the better of the class, so's Jonathan and Sylvia, and they sometimes call us to . . . work on the computers and find codes for the next lesson or something that we have to work out."

In reply to my question asking how they feel about setting next year, Carol referred once again to the group of eight, in which she sat next to Nicola:

> . . . because sometimes Sir is teaching the front of the class something and the eight people at the back have got to do something out of their books different. It'd just be easier for everybody to be being taught the same thing.

Similarly, Jake argues that:

> I think it's an *extremely* good idea . . . because you wouldn't want somebody to be too, have work too hard for them if they didn't want. And if I needed harder work then I'd want to have it.

Lying behind their advocacy of setting was a fairly fundamental fixed ability belief and of course their experience of the emphasis on speed and the competitiveness that goes with it, a crucial element in the self-perpetuating distinction between sets. For example, David describes himself as totally committed to doing things as fast as possible, particularly when he is working with someone else: "If I'm (racing through it) I'm normally working with somebody else and we're working them out together and we do them dead fast."

John is one of the lower-attaining students, and he too is in favor of setting. This is interesting, because he is at pains to tell me that his teacher makes a deliberate effort not to exclude any pupil from classroom exchanges "so they don't feel left out." Despite this apparent skill on the part of this teacher in handling the mixed group, John says that setting will be good because "I expect to be, like, in a group with, like, the people who are the same because then you're, like, you're all, like the same and someone, like, won't get, like, left behind and have to catch up." It is the comparison with others that is associated with speed that is the problem: "sometimes, like, some of them they don't work as, like, fast and they don't get as much done and they feel, like, bad about themselves because they haven't done as much work."

Jake presents himself as an exception to the speed issue, however, arguing that getting things right is more important to him. When I ask him about boys being particularly keen on speed, he tells me that his friends try to be fast—one did a test in a minute, he claims—but they don't do very well. But he wants the marks, and tries to score 100%:

> I try to. Most subjects I don't get exactly one hundred percent. In maths in my most recent test, it was a practice one, I got ninety-nine point five . . .

I comment that this is very good, but he feels the need to justify this loss with a lengthy story about this being due to a slip rather than lack of understanding— he forgot to work out the problem in the required way: "so I lost five. It was because I forgot to work it out in a particular way but I did get it right." He is quite persistent about this issue—when I try to move on he says: "it's better getting the answer right, right and not working it out the same way, as you not working it out at all." This insistence that he would/could/should have gained 100% seems crucial to his taking up of an identity of someone who does not subscribe to the demand for speed: to sustain this he needs to demonstrate that he is extremely good at mathematics—good enough to be above the speed issue—and this makes sense of his apparently self-contradictory claim later in the interview that he is not competitive:

> . . . in Year 3, I wanted to be better than everybody else. But it isn't that I want to be better any more it's that I want to . . . learn and do well for me. . . . I don't need to do well for somebody else. I want to improve, I don't want to go down.

Epistemologies of Mathematics

These discourses of mathematical ability and learning appear to be connected to the way in which the students perceived mathematics as a subject: they generally described it as a collection of disconnected topics, to be learned with the aim of procedural mastery to produce answers which were right or wrong. This perception was fed by their experience of high-stakes assessment in which mathematics becomes very risky, prey to luck or memory. Becky, who describes herself as "in the middle" of the class in terms of attainment, explains how she was nervous during the previous year's SATs because she is never sure that she will be able to do mathematics tests, even though she is reasonably good at it:

> [*Did you feel anxious about the SATS?*] Slightly, because in some of the practice papers we did every now and again I'd get a paper that we'd do and I'd do quite well and then other times I got a low mark . . . it depended on the questions and what type of questions I got as to what level I would get. . . . I was a bit, I had a bit of butterflies just in case the paper, I found the paper really hard . . . it was like the luck of the draw with what paper.

Carol, who is one of the group of eight, described herself as a very competitive person who likes to do well and get right answers, which is why mathematics is her favorite subject at the moment:

> Because maths has got straight answers and [in] English there's lots of different answers that you can get but maths is simple and there's only one right answer so I think it's simple in that way.

However, she tends to rely on memory because she does not always really understand and it is this aspect of mathematics that increases risk:

> I enjoy doing problems but sometimes . . . I find them hard to figure out because . . . I look at it and can't understand it sometimes . . . I think it's more about remembering things. Once you've been taught them once you've just got to remember them and then you can do most things.

Lack of understanding cannot always be replaced by memory, of course, and Year 7 algebra was proving a challenge for some students, engendering some strong feelings about abstract versus concrete. One of the first things that John says in the interview is that he doesn't like algebra, because he can't understand it. Faced with "$2x = 6$," he says "I wouldn't have a clue what you were on about"— his reaction on seeing "x" in an exercise is "oh no." When we explore this a bit more, it transpires that, while he is happy enough with "$x = 6$," he is confused about what "$2x$" means—when I explain that it means 2 *times x*, he says "I thought it meant like 2 add x equals . . ." This is the kind of thing that he forgets, he tells me, and doing algebra would only be likeable "if I like could just do it like off by heart." David has similar problems with algebra, even though he is, he tells me, above average at mathematics (although he qualifies this slightly, saying that he gets "a bit distracted" in class) and that it is his favorite subject:

I don't like that [algebra] because I can't really like fill them in because they're like 7*a* add thingy add thingy because they're normally formulas aren't they . . . and I can't do that . . . I just don't get the numbers and the letters . . . things like 7*a* add 8*b* add 2*c* or something. I don't get adding the other letters together.

Algebra difficulties notwithstanding, what makes him good at mathematics in his view is his memory: "I've got a good memory and I can remember things and like methods, how to do it."

There is, however, another mathematics life which pupils described as something of a bonus (i.e., not part of the usual mathematics curriculum) for good work: they sometimes did "investigations." Despite her emphasis on right answers, Carol likes investigations because they require more thought than the basic exercises which take up most of their lessons:

You're learning how to figure out things in a different way . . . When you're doing adding and subtracting you're just, like, figuring out things in an easy way but you've got to look at it in a different way in investigations and things like that . . . you've got to figure it, it's not, like, straight and simple and you can't just say "ah well you've got to do that". You've got to kind of figure out what you've got to do before you do it.

Investigations present an opportunity for a greater sense of ownership and engagement in mathematics, and a corresponding shift in the associated personal epistemologies away from that-which-must-be-memorized-or-lost and towards something-which-can-be-built. However, investigations were not described as an integral part of learning mathematics, and they tended to be an activity associated with the higher attainers. Even for these students, investigations were a "fun" bolt-on, a departure from the norm often generated by external bodies in local or national competitions. Their potential benefits were thus minimized, showing very little impact on the pupils' overall views of mathematics. Nicola and Jake were thus somewhat unusual in describing mathematics as potentially creative in a general sense outside of investigations. Nicola contrasted mathematics and English in a way which was the complete opposite of Carol's explanation as she tried to describe the complexity of inter-relationships in mathematics:

Maths is like a lot more . . . complicated thinking than . . . science or English because . . . well, in English there's only a few . . . sentences and things like that, whereas maths is a whole . . . different thing and a whole variety of things to be introduced to . . . I think there's lots of different subjects but in the subjects there are lots of different things that . . . come back together. So one subject, say algebra, there's actually different topics in that . . . range but they . . . somehow fit in together.

Trying to explain the benefits of working in a group, Jake also described doing mathematics as creative, emphasizing its *lack* of "straight answers":

> In maths you can have opinions because in addition you can't really have an opinion but you can have opinions for different things. Because if it was like a puzzle or something you could have an opinion because it could be done two different ways or maybe they might spot something you haven't spotted and you say one thing and they say the other. And so then if you like work together you can put two together and get a stronger answer.

Year 7 Mathematics Identities

Like the Year 5 children in Chapter 2, the Year 7 students organize their mathematics experience in terms of comparison with others, but their grouping of themselves and their assessment of who is "good at maths" is more defined. While they have lost some of the competitiveness of the Year 5 children, they adopt the language and assumptions of setting unquestioningly, and the emphasis on speed remains as a marker of ability, among the boys at least. There are also suggestions that gender is an organizer in the girls' narrativizations: they draw contrasts between themselves and boys in terms of their behavior, portraying boys as more confident and less compliant than themselves. Nicola complained that the boys—even if they are good at mathematics—"mess about":

> Out of the eight people who are in the like higher group in our lesson there's only two boys and the boys normally tend to mess about a bit. They're the ones that are normally, like, talking and they don't listen as much. Certainly at primary school, they always sat on a table further apart because they didn't do the work . . .

Carol tells me that the boys in the higher attaining group are not very hard-working, preferring to laze around if they can:

> . . . they're quite slow and just get on at the speed they want to. Sometimes they can't be bothered doing the work so they just sit there . . . I don't think they're that bothered about what score they get. They just, it's not their favorite subject and they want to do better in other subjects so they're not bothered about maths.

Sylvia also comments on the boys' behavior: although she rejects the idea that there is any consistent difference between boys and girls in terms of how they do mathematics, she goes on to explain other differences in their public behavior which are reminiscent of Baxter's research:

I think the boys aren't usually afraid to put their hands up and the girls will usually sit there quiet but . . . the girls tend to just sit there and watch. . . . I'm like that though, I don't like putting my hand up that much [*because you're worried about looking stupid or . . . what?*] yeah, that's it basically (*laughs*) . . . they're always joking about looking silly anyway so it doesn't usually matter as much to them.

In terms of their relationships with mathematics, even the more successful pupils described themselves as largely prey to luck or memory, casting aside their professed beliefs in effort. Nicola appears to have more of an identity of engagement than other pupils, matching her more creative epistemology of mathematics. She prefers to work out her own methods if she can, because it gives her a greater sense of ownership: "I think it's just because you actually know a bit more why you're doing that method . . . because then we get a sense of trying to do it all ourselves instead of getting part of a method . . ." However, like the majority of pupils she is also subject to the pressure of the competitive environment that prevails in the classroom:

I think it's a bit hard sometimes because if you don't understand a lot of it it's a bit embarrassing sometimes if you put your hand up and say "I don't get it" . . . If we have a test and everyone else gets a higher level than me I do feel a bit . . . embarrassed. . . . it makes you want to . . . work harder.

As I have noted in Chapter 2, ascribing one's success to luck or "just hard work" rather than (true) ability is a recognized ploy which may be used more by girls, whereas boys are more likely to attribute failure to luck and success to (natural) ability (see also Dweck, 1986). Indeed, the boys were more likely to claim to be good at mathematics (even if they were not among the higher attainers, as in David's case) and to project a sense of confidence, in line with the research reviewed in Chapter 2 (see also Reyes, 1984). Jake for example described himself as "quite a strong mathematician, probably because I like working out things." Jonathan presents a rather extreme case of this: he is "gifted and talented," so he tells me during the course of the interview. He presents himself as someone who can handle other people (when I ask if he will be in trouble as our interview is taking much longer than scheduled, he says "no one ever shouts at me. I can talk my way out of it"), and he positions himself as an eccentric, into calculators, codes, computers and hacking, with extreme opinions. In a similar vein, he emphasizes his difference from other people—including the teachers—in terms of his approach to mathematics which he portrays as distinctly individualistic:

I approach maths in a very different way to a lot of people. I do it, I tend to use more tricky methods which doesn't always work out right but that's how I like to do it . . . I don't usually like conform to what the teachers say to do because I'll do what I want because, I know that sounds stupid but I do what I want because I've done it before and it's worked out right.

He relates this difference partly to his personality, which he presents as somewhat unstable, invoking the images of (male) mathematicians observed by Mendick (2005a, 2006): "Usually I'm very commanding. I like, I take over a group. I have a lot of mood swings and it all depends on what kind of mood, really, how I react really." While Jonathan describes himself as doing things differently from the teachers almost on principle, the majority of pupils described themselves in rather passive relationships with teachers. However, Jake appears to have more of a negotiated relationship with his mathematics teacher, and reflects for much of his interview on different relationships in different subject areas: he makes the insightful observation that a teacher needs to see where the learner is coming from in terms of what they already understand and their need to know the purpose of what they are learning about:

It's good that our teacher explains it a lot, he tells us what we're doing, because in different lessons sometimes I'm left wondering . . . Some subjects I find very difficult though . . . I find geography difficult because of our teacher. She doesn't ever understand where I'm looking from so then I never learn anything.

Some of the higher performing students described background features which suggested that they had important advantages which helped them. Nicola benefited greatly from support at home: she described how her project engineer father and her brother were both good at mathematics and could explain it if she didn't understand. Her father "does maths for fun," doing mathematics examination papers alongside his children. Carol had similar support from her father, an adult IT teacher, and Jake's father also utilized his computing expertise to help:

The good thing is if we had a maths, English or maths or something test and it said revise these things then what my dad would do or something is get on the computer—he did this all for me—he got on the computer and invented a game for me to play.

Just as he has done when the question of boys versus girls comes up ("when you look at all the chess champions and intellectual champions they're always boys. . . . it sounds very, very sexist but there's hardly a girl amongst the intellectual ranks of the world"), Jonathan gives voice to opinions about social class which are part of a publicly less acceptable discourse of difference. He is, he says "ridiculously competitive" and that means "being socially different. If you want to be clever . . . I wouldn't mind being, like, fitting in with the crowd but then I see what the crowd is like." Children who do less well at school, he says, come from backgrounds which do not support education:

> Some people are brought up like playing football until half past nine every night and they don't have any time for school. So it, they tend to drop behind . . . And it sounds, like, really snobby but today's youth they don't sit down and talk to you, they're more, like, out playing football or whatever.

It is noteworthy perhaps that Jonathan does not describe getting help *per se* for school work; rather, he invokes discourses of class and culture to account, implicitly, for his success. His own background features a father who has passed on his ability to tinker with computers:

> I've always been interested in computers, ever since, like, my dad used to . . . to take apart and put together computers. . . . if the computer broke he had it all over the bathroom carpet in like fifteen minutes and he'd have it back together working better than perfect in around three minutes. And he showed me how to do it and I can take apart our computer and put it together again in working order. And then I realized computers weren't just cogs, it was maths . . .

How can we sum up what the Year 7 group have said about doing mathematics? It is possible to see a number of interconnecting strands that make up their mathematical lives so far, sometimes from defining moments such as their experience of Year 6 SATs or from more general background features such as parental support, particularly when exams have come to the fore and more generally "coming from a mathsy family." We can also see how these students take up the positions offered within the practice of setting and its related assumptions about ability, speed and the need for a different, less challenging, curriculum for some students, advocating it strongly whatever set they are destined for. A further thread which runs though their narratives concerns the role of gender in classrooms, in particular boys' behavior and attitudes as presented by the girls—arguably, this tells us more about the girls than about the boys in terms of their self-positioning as "the sensible ones" who get on with the work. Their portrayal of the higher attaining boys as frequently "cruising" rather than working fits of course with the discourses uncovered by Walkerdine's (1998) work, noted in Chapter 2.

The students' relationships with mathematics itself appear to be colored by the institutional structure in that it supports a sequential view of mathematics—connectedness is unusual. Building on the Year 5 data and the research reviewed in Chapter 2, we can speculate that where this occurs it may be sustained both by resources which come from outside the classroom, and by the different experiences that being a member of the "group of eight" affords.

Learning in Ability Groups: The Year 9 and 10 Stories

By Year 9, setting was firmly established, and top set pupils described a very different experience and perception of learning mathematics when compared with those in lower sets. Setting for the Year 10 pupils had an extra dimension in terms of being explicitly linked to Key Stage 3 SATs results, and the sample included three pupils who said that they had been moved to different sets as a consequence. The polarization of experience suggested by the Year 7 accounts continued and was extended within the context of impending GCSE, to be taken earlier than the others by two of the pupils in the top set, and at different levels of difficulty (tiers) between the top and lower sets. Most notably, investigation appeared to have dropped out of the curriculum, except for its role in GCSE coursework, when it was more likely to be described as an individual rather than a group activity. The most striking development, however, was the extension of the differences in self-positioning between boys and girls which were already apparent in the earlier years and which are best illustrated in the accounts of the top set girls.

Top Set: Living Out the "Ability Group" Discourse

Year 9 and 10 students made the same basic claims about effort and ability as Year 7 pupils, and they too invoked natural talent when discussing mathematics learning as they experienced it. All of the pupils in the sample were positive about setting, most citing the now familiar argument that lower attaining pupils are put under pressure in mixed sets, and some, especially girls, arguing that it is easier to get on with work in sets (because lower sets, and especially boys in lower sets, "muck about"). Year 9 Michael raised the other central issue in setting: the importance of speed. Although he conceded that effort can achieve results, he argued that mathematics ability was innate and clearly evidenced by speed of work. This was why he had himself been selected for the top set:

> . . . it's just like the rate of work when we were all kind of mixed. Probably a bit of natural talent as well because some people can just like . . . see the answer in their head sort of thing so people like that, if they can just look at it and work it out it saves a lot of time so . . . because you've got to be born with like, some people are good at English, some people are good at maths.

Michael judges that "catching up" would be a question not only of working hard but also being able to survive at the top set pace. A slightly different analysis of the differences is given by Year 9 Luke: although there is a general emphasis on speed in the class which he acknowledges puts pressure on him, he resists this, describing himself as different from the others because he is able to see things in mathematical perspective:

> . . . I can understand how it works and I can see how it works and when I take time and think about it and look at it I can see it and I can do it right and correctly. . . . Whereas [others] . . . might be thinking "Oh, what's this?" and I'm "Oh yeah, that's that and that and that".

Luke's self-positioning here as someone who is good enough at mathematics to prioritize understanding over speed is similar to that of Year 7 Jake. He goes on to describe making connections and creativity in mathematics:

> It's good to be able to do that because, say, in between algebra and doing some normal maths or between algebra and say constructions and things like that, you can think "Oh yeah, I've done something a bit like this so I can use that knowledge to help me with this", and then maybe the two things actually combine together to make a different thing and when you actually do that different thing you think "Oh, this is just this and this put together which is easy so I can do that".

Luke argues that what distinguishes him from other pupils who are less successful at mathematics is "the difference between being able to do maths and being able to do the maths investigations," thus defining the parameters of a participative identity which is echoed by other pupils in the sample. Indeed, the top set students describe an emphasis on understanding and challenge—for example, Michael says that top set pupils are more likely to put their hands up to ask for explanation because they are more concerned with understanding, while Year 9 Rosie says that there is more intellectual challenge:

> I like it because the work is more challenging. . . . Stuff we did when we were mixed abilities, it weren't as challenging as what we get now so I like it more because it gives you more of a challenge to try and get the answer.

Making Mathematical Connections: Negotiated Learning

Top set pupils reported that they experienced distinctive mathematics lessons in terms of both content and relationships with their teachers, and they described their relationships with mathematics as different from those in the lower sets. Daniel, in Year 9, describes the top set as providing the opportunity to make connections and so to see and understand mathematics differently:

> . . . we do more of, little bits of more things whereas the people who are lower down do more things with little bits so they don't see as much . . .

we sort of see it, we sort of see *all* the maths problems and how they connect to each other and we understand it more.

A number of pupils were working on an external competition which involved sustained investigation work in teams. All were enthusiastic about this activity, partly because it involved working with others, and partly because it entailed problem solving and taking a broader and creative view of mathematics. Thus Year 9 Georgia described an epistemology of mathematics which emphasized the esoteric aspects of the subject. She finds mathematics the easiest subject at school but her favorite is art, and she notes the similarities between them with respect to opportunities for creativity and self-expression:

> Well, like, sometimes when you're doing, like, certain course work in maths you get to, like, use your own sort of ideas, like, and, like, as you do in art as well . . . Like, if I just, like, get the outline of the investigation then, like, you just can put, like, whatever to it yourself. . . . It won't be the same as anybody else's idea. You get to add, like, a part of you, like, into the project or whatever you're doing . . .

Despite the pressure to perform and compete which comes with GCSEs, the Year 10 top set pupils maintained an emphasis on working together to understand mathematics. Christopher described how his investigation experience working alongside Harry had been beneficial and creative:

> He did one bit, I did another bit, we kind of put our own ideas together . . . because sometimes you look at it one way and they look at it another way and then you can, like, work together with the different ideas you've got.

Christopher says that he was "champion problem solver" in a previous school, and that he has enjoyed being in the top set because he has had an opportunity to explore mathematics more, to understand a problem and its solution rather than following algorithms—"it's using the stuff that we've been taught at school and applying it to the question." Being in the top set means more time for this, he says, justifying setting once more:

> Because we're just not, it's not like we're having to wait for the other groups to catch up with us because we've nearly, like, finished the syllabus kind of thing so we've got that freedom and just go back and do the things that we need to learn.

The benefits of the top set are not merely a question of getting through the basic material in order to move on to more creative activities, however. Several students reported a change of relationship with the teacher towards acting as a resource, as Michael describes:

> [The teacher's role is] mainly just trying to explain things . . . Once we get rolling we're usually quite independent and we'll, once we've checked our answers and then she'll write or we'll run it through her just to make sure she thinks we've gone about the right way of doing it. But that's about it really . . . I only ask her as a last resort. I usually ask the people around me first.

Georgia says that she sometimes tries to work out her own methods, and sometimes finds that these are easier than her teacher's suggestions. But she clearly sees the teacher–pupil relationship as one which works best with negotiation on both sides:

> I think you've got to, like, get on with your teacher, try and see things from their point of view, like, when you're trying to work something out which makes it easier. . . . [You have to see], like, how they're trying to work it out and show you how to work it out. . . . you're using like what you already know and then, like, adding some bits that maybe you didn't know, like, with the teacher's help or whatever . . .

Year 10 Kate had been moved up "from the bottom" following a good performance in the SATs in the previous year. As a result of her experience, Kate believes strongly in effort, but she puts her previous lack of application down to the teaching she received in lower sets:

> The teacher [is] not very strict, she just sits there, tells us what to do and then just leaves us to do it. She doesn't help us or anything. . . . That's one of the reasons that I got moved up because Mr Philips saw that I could do it so he took me into his class and he goes through it all. . . . She'd just sit there even though you'd asked her. She'd say, "Oh well, you know it all so just do it." And Mr Philips he'll go through it, he'll come over to you, put attention into you and actually talk you through all the bits you don't understand.

Lower Set: "Lower Ability" and Lower Expectations

The lower set students told very different stories about doing mathematics. Their accounts suggested differences in classroom cultures and curriculum, and in their perceptions of the nature of mathematics, which were reminiscent of the commonsense applications stressed in the low level texts studied by Dowling (1998, 2001), which I will discuss in Chapter 7. These pupils presented an account of mathematics which suggested marginalization rather than participation, describing it, as Schoenfeld (1992) does, as something "done to them" within a climate of performance and "getting by" as isolated learners.

As I have already observed, all the Year 9 and 10 pupils were in favor of ability grouping, usually on the basis of fixed ability beliefs. Year 9 Lizzie, herself in a lower set, argued that it should be retained, invoking the sequential view of mathematics:

> Because it's not fair because if there's people, like, lower ability people [in the same set] the top people can't reach their full potential because they're having to do lower work that they already find easy.

The lower and higher sets were most obviously different in their responses to investigations in mathematics. An extreme example of this difference is provided by Year 9 Trevor, who does not in fact appear to know what I mean when I ask him about whether he likes doing investigations. He describes mathematics in terms of performance and memory, and he has an instrumental view of its use: he wants to be a truck driver and is focused on learning the school mathematics necessary for that job and life in general:

> I want to be a truck driver so I've got to, like, see how many hours I've done . . . work out the exact mileage and everything. . . . When I go with my dad and my mum shopping, like, buying stuff and it's seventeen point five per cent, they might need to work it out before they go up and buy it . . .

Commensurate with his emphasis on memory is Trevor's basic concern with his marks—he talks about his performance often, in sharp contrast to top set Luke, for instance, who never mentions performance directly. When I ask him about how he did in the recent SATs, he gives me a lot of figures, which in fact bear no relation to the actual reporting of SATs results, which are given in terms of numerical levels, the average for Year 9 being Level 6:

> I'd say I'd got about . . . over fifty per cent, like the average, around fifty-eight-ish, something like that. Because the last one I got overall I think was seventy-eight per cent, something sixty-ish . . . so I was pleased with that because my mum said she was happy. At least I got over fifty per cent.

Lizzie is in the same set as Trevor, and she also sees having a good memory as an essential requirement for being good at mathematics because sometimes it must replace understanding:

> You've obviously got to understand it but some of the formulas, they don't really make sense, they're hard to explain why they do it, it just works. So,

but it is hard because there's so many different formulas you've got to remember. You've got to have a good memory.

What is interesting about Lizzie, however, is her obvious liking for investigations, and in this respect she is somewhat more reflective about what is involved in mathematics than Trevor. She contrasts the drill of set exercises and the creativity of investigations in a way which is similar to Luke's:

It's not like [you] have a sum set for you. You can start working it out and then you'll get different leads and you'll go off in a different direction and everyone at the end of it will come out with something different. . . . With a set sum you just have to get the answer but with an investigation you can answer loads of questions and you can try it with different things.

Lizzie's particular mathematics history is that she used to be in the top set, but has been moved down. Despite her obvious liking for the challenge of investigations—which she had undertaken in the top set in Year 8 but not in her Year 9 set—she lacks Luke's confidence and self-positioning as a participant, falling back on an epistemology of mathematics which is based on fixed ability. Like many other students, she believes that "some people have just got it and other people just find it harder," and that being good at mathematics is simply a sign of cleverness which has social exchange value:

The teachers say you'll need maths when you're older. . . . But I think it's just to prove how clever you are sort of thing.

Year 10 Paul also says that he enjoys investigations but this is because they allow him to work at his own speed and to evolve his own, more memorable, methods. Unlike the top set students, his preference is for working alone: working with others "puts him off." This appears to have quite lot to do with the competition created by the upcoming GCSEs and its interpretation in the lower sets in terms of performance rather than the mastery of the top set. Thus Ben, also in Year 10, was highly competitive about his GCSE practice investigation on "noughts and crosses" (Tic Tac Toe), with limited benefits for his mathematics understanding, it seems, in that "You want to get to the lesson and just get into it . . . and you're wanting to find more than other people." His reply when asked what he had learned from the investigation is simply "how to get on" in terms of sticking at a task. What he might have gained in more mathematical terms is unclear, despite much prompting:

I didn't hardly speak at all, I did a lot of work in that [*And what was the end point?*] There were lots of different ways (*pause*) can't quite remember [*What was your target?*] As many as possible and trying different grids and how many different, enlarging the grids [*Did you actually generate any kind of formula for how things related to each other?*] No I don't think so. [*So what did you learn maths-wise?*] Puzzles sort of thing (*pause*) [*Is there anything you've learnt from it that you'd be able to apply next time?*] (*pause*) I don't think so, no.

"Maths is Boring": Ritual and Marginalization

Lower set experiences of doing mathematics at school seem to be largely colored by a perception of the teacher–pupil relationship as one based on a positional teacher authority, an outlook which fits with an emphasis on performance. Trevor, for example, appears to lack the "pedagogic awareness" (Black, 2002a) which many top set pupils have of learning as a process of negotiation between teacher and learner: asked if he ever shares his own ways of solving problems with his teacher, he responds in terms which suggest that the main aim is not to get into trouble: "If we've got the right idea but don't get the right answer, they don't tell us off, still like, at least we've tried." Lizzie's ideal mathematics lesson suggests that lower set teaching provides few opportunities for engagement:

> I'd make it more interesting. . . . Because we get exercises and it does help you to understand what you don't know and what you do and get to work on it but I wish they could think of new ways to teach it ... more investigations so you could sit there and do it for yourself ... and make you think properly.

Year 10 Anna also described how she likes to be active in the lesson, to think for herself and to work in groups because this is more fun. Her experience of mathematics lessons, however, is that they are a matter of endurance: boring, pointless and lacking in activity and fun. Enjoyment is a central theme in Anna's interview: she sees enjoying mathematics as the main difference between pupils in the top set and herself, and she is emphatic that she can never be good at mathematics because she does not enjoy it, although she concedes that she might begin to if she could do more active investigations "because we're actually doing it ourselves and not having to listen to someone explain it." Her perception of lack of control over what she does in mathematics is crucial:

> It depends if you want to do it or not . . . if you choose to do it you enjoy it more because it's what you want to do. But if you don't chose it and you get forced to do it then it's different . . . we do pointless things all the time.

Tom is also quite disaffected with mathematics, and although he describes himself as competitive, he claims that school results are not a major concern for him: "if I get a bad result I'm not just gonna be really that gutted really. I'm not that bothered." Like Ben, he describes working alone on his Year 10 GCSE practice investigation, and he too cannot remember what he did or what he learned. He finds mathematics "boring," a response he provides in answer to my question about typical mathematics lessons, which are a matter of "Just go in and he gives the textbook out and do questions out of it." Despite his presentation of himself as disinterested, he shows some animation in response to my question about what he would change, describing mathematics lessons as stifling his creativity:

> ... do more practical stuff and, or maybe do some tasks instead of just out of the book and messing about doing stuff out of the textbooks ... [*How does maths compare to other lessons that you do?*] ... you get to do stuff of your own like, you can do stuff and like you don't have to copy out of textbooks. You can do like poems and make up your own stuff or say, make your own stories, stuff like that, not just copying out of books [*So it's more creative?*] yeah.

Taking a Position: Interactions Between Discourse and Experience

The Year 9 and 10 students' accounts suggest contrasting epistemologies of mathematics and experiences of teaching and learning between the top and lower sets which are associated with corresponding identities of participation and marginalization. Top set learning appears to be characterized by Wenger's prescribed mix of opportunities for engagement, imagination and alignment as pupils describe themselves as not only learning the basics but also how to manipulate these in a negotiated and reflective mathematics. Lower set learning on the other hand appears to be a simple question of blind alignment. Thus Georgia, Luke, Michael, Daniel, Christopher, and Harry describe teaching and learning relationships in their top sets in terms of their own participative identities, while lower set pupils Anna, Lizzie, Ben, Paul, and Trevor describe their lessons as dominated by memorizing facts and algorithms in the context of relationships with mathematics which extend for some to identities of marginalization.

The data do not fall uniformly into such neat categories, however. Among the top set students, Georgia seems to be the only girl to express a positive mathematical identity in her description of herself as someone who can do, and enjoys, mathematics. For example, Jenny, one of the Year 9 top set girls, describes her experience of mathematics in ways which match more closely to lower set marginalized identities:

> The teachers tend to show the hard way . . . a lot of the time. They do show you an easier way but only briefly because they just want you to do the complicated way so you probably can pick up more marks or something.

Jenny is less happy than the others in the top set, and she raises the same issue as the Year 7 girls about the boys' behavior, criticizing them for falling into a habit of emphasizing speed:

> Boys just scribble it down and I don't think they really care what happens with it . . . "try and get more done", quantity not quality. . . . there is more lads than lasses that go faster and . . . the hand-writing is like dead scruffy and you can't read it. But they, like, go dead fast.

As in the case of the Year 7 girls, Jenny's comparison is perhaps indicative of her own particular self-positioning with respect to mathematics in the context of the equation of speed with ability that prevails in the top set. She compares herself with Daniel, who she frequently works with, in just these terms:

> I think he just picks it up better than me. Once we've both got it we're probably the same but he just like picks it up and once he's explained it me then I get it.

Other top set girls also tended to describe themselves as less competent than they might be. Despite her advocacy of the challenge of top set work, Year 9 Rosie expresses strong likes and dislikes for particular areas of mathematics, avoiding any which she finds difficult, particularly algebra. When I question whether this isn't a challenge too, she replies that "There's just too many things to remember about it." In Year 10, too, the girls appeared more vulnerable to self-positioning in terms of the opinions and actions of others. Kate, for example, explained how nervous she became over exams, despite the fact that she had been moved up to the top set on the basis of the previous year's performance:

> If you do rubbish then your parents might have gonna think you're rubbish and you're not gonna do very well in your GCSE year. So a lot of pressure on you. . . . They're always saying, like, I'm not trying too hard but you are but they don't know how hard you actually are trying.

She was also anxious not to get things wrong in class, because "[if] it's miles out then people will laugh at you but not in a horrible way, joke about it, and you feel a bit down because you got it wrong"; unusually for the top set, she preferred not to work with others because this meant that she didn't have to challenge and/or defend her ideas:

> I tend to work on my own . . . because you've just got your own freedom and that and whatever you say or you think in your head so you don't have to challenge what other people say.

Rachel had also been promoted to the top set at the end of Year 9 on the basis of her SATs, and displayed a high degree of creative engagement in mathematics. She maintained nevertheless that she was no good at mathematics and did not like it, despite her insightful treatment of the noughts and crosses investigation which meant so little to Ben. Here she explains what she means when she says she does not like mathematics:

> Maths, I don't really like it because I don't see the point of it. I like ... I don't know ... Like when we're doing work, all the algebra things, I think "What is x and what is n? Why are we trying to make that y, what's that all about?" I can't understand why I'm doing it so I can't really understand how to do it.

On further discussion about this it transpires that Rachel is in fact able to manipulate formulae but cannot tolerate not knowing exactly why the mathematics—quadratic equations for instance—works out or what it signifies. She demands a high level of understanding which she cannot always have. Thus her problem with mathematics "not having a point" is not like Anna's complaint; she wants to knows more, not less, and she is prepared to work problems through to get this. In terms of her personal epistemology, Rachel is an engaged and creative mathematics student, but her tendency to interpret her desire to understand and connect as weakness seems to present a major obstacle in terms of the development of a participative identity of being able to do mathematics.

This resistance to a positioning of self as good at mathematics is particularly illustrated in the case of Sue, one of only two students—Harry and Sue—in the whole year group to be taking GCSE mathematics early, in Year 10. While Harry describes himself as "above average," and talks enthusiastically about real-world applications of mathematics and his ambitious approach to the coursework noughts and crosses investigation, Sue describes herself modestly as "quite good" at mathematics, which is "okay sometimes": her almost apologetic approach to the early GCSE is that it means that she could take it again if she gets "a bad result." She finds the coursework stressful, telling me that she didn't particularly enjoy it because of the deadlines "and you don't really know where to start sort of thing." This effect was made worse for Sue by the fact that she and Harry were the only pupils undertaking the coursework—she would feel "more confident," she says if everyone was doing it. The fact that Harry was with her in this enterprise did not help, because he was, in her view, much more advanced than herself:

> You know, Harry's a very good mathematician so his [coursework] is really good. . . . he is more advanced, he knows more things that we have to do . . . in tests he can take the formulas out of the front of the paper and put

them to the questions and some of them I don't know what to do with them.

Sue concedes that some of these questions covered issues that they had not yet been taught and that Harry "must do things at home." She agreed—but very reluctantly—that in the right circumstances she could be as good as Harry. It was not clear, however, if Sue really believed this; asked directly about effort and ability, she maintained that effort was important, but then, as many other pupils did, went on to say that "Some people really aren't . . . as good. Some people can't learn as much."

There is some indication, then, that gender is crucial in the development of identities of engagement and ownership. While girls such as Sue, Lizzie, Rosie, Jenny, and Rachel show clear indications of making important connections in mathematics and a taste for investigative mathematics, they lack identities of participation and "being good at maths" such as those which are particularly evident in the accounts of the top set boys. A neo-Vygotskian perspective such as Edwards and Mercer's (1987) or Mercer's (1995) analyses of classroom dynamics, suggests that the kind of learning relationships which the top set report should be of benefit to all pupils. On one level this is indeed the case: the top set pupils, including the girls, expressed epistemologies of mathematics based on understanding, making connections and creativity. Yet the girls' marginalized identities present an anomaly which is also recorded in earlier research showing top set girls' anxiety as I have noted in Chapter 2. As Stetsenko and Arievitch (2004, pp. 478–9) suggest, the equation of subjectivity with participation in a community of practice overlooks "how particular selves are produced, or [. . .] the active role that the self might play in the production of discourse, community and society itself." All this raises the question, reminiscent of Lemke's observation on the uniqueness and typicality of classrooms, which is identified by Sfard and Prusak (2005, p. 14) as central to an investigation of learning: "Why do different individuals act differently in the same situations? And why, differences notwithstanding, do different individuals' actions often reveal a distinct family resemblance?"

Sfard and Prusak's answer to this question is that "it is our *vision* of our own or other people's experiences, and not the experiences as such, that constitutes identities. Rather than viewing identities as entities residing in the world itself, our narrative definition presents them as *discursive counterparts* of one's lived experiences" (p. 17). If we consider Gee's (2001, p. 111) notion of unique trajectories through "Discourse space" we can understand how these pupils might position themselves and be positioned within the mathematics classroom, and the different ways of being that it affords, so that identities are formed over time and have their roots in repeated positionings and repeated stories. Thus what pupils experience in mathematics classrooms will always be colored by the stories they tell. Their narratives frequently appeal to a Nature-identity (if we include the idea of natural ability in this), but also incorporate Discourse-identities which draw on discourses about mathematics and of gender differentiation. Institution-identity is clearly ascribed in terms of "ability group" membership, while Affinity-identity appears in the students' accounts of their within-class groups, such as the gender groupings which girls rather than boys allude to frequently, or the "group of eight" in Year 7.

It is not clear what the origins of the differences between boys and girls observed here are, but we can speculate on the possible effect of the emphasis on speed that the girls frequently comment on, its institutionally-sanctioned equation with success in mathematics, and its association with the boys. We have already seen how the Year 7 girls talked about the boys in terms of their unwillingness to comply, and their confidence. Year 9 Lizzie also has quite a lot to say about boys' behavior, calling upon mixed discourses of gender differentiation, women's resistance and equity:

They're louder. They *have* to get it right, they have to shout it out first. They're more gobby. They have to get there first like to prove that they're there . . . because they think men are more dominant. That's what they think but I think girls are cleverer, they might be quieter about it but they still get there. I think everyone works it out the same, everyone's okay in my thing, we're sort of all on the same level and all together but boys are like, if you don't, if a girl doesn't understand it lads are like, "God, you're well thick", sort of thing.

When I express some shock at this, she qualifies her account slightly, but she still persists with an account of gender differentiation which emphasizes the boys' competitive behavior, taking this opportunity also to underline the value of the sets in protecting students from within-class competition:

They used to in junior school but [now] everyone's just sort of on the same level so it's all right here . . . Because we're in our groups everyone's on the same level so everyone sort of doesn't understand what everyone else doesn't . . . Girls . . . don't really mind admitting that they're wrong and that they don't understand it. It's not that bad now because like we're in Year 9 . . . everyone's sort of like maturer now, they'll admit it if they don't understand it but usually they, it's the girls that'd admit it and boys wouldn't . . .

Lizzie's story invokes a long history of dealing with boys. Taken alongside the research reviewed in Chapter 2, and the Year 5 data, we can speculate on her story and that of the other students as products of repeated positionings in mathematics classrooms over the years but also of the dominant discourses that frame those experiences. Black's data suggests that girls have less opportunity to talk themselves into acting as participants—they are always on the margins, or feel themselves to be so. In addition, though, the discourses of ability, competition, performance and comparison which are perpetuated by constant high-stakes testing and its use in the audit culture appear to have considerable impact on what happens in classrooms and how it is constructed by both teachers and pupils. In particular, the advocacy of the concept of "ability grouping" and its related sequential view of mathematics in so many schools has a complex role in pupil identities. All students without exception were supportive of setting as a principle, but their defense of this practice frequently drew on an argument which only has real currency in the climate created by SATs and GCSEs—group membership correlates with the speed of delivery of the material, driven by the demands of coverage rather than understanding. Despite their professed belief in effort, set membership plays a part in pupils' identities, not only in terms of self-perception as "good" or "bad" at mathematics but also in terms of their experience of mathematics and the available positionings that membership of a particular group or sub-group brings.

4
Moving On and Moving Up?
Entering the World of the Undergraduate Mathematics Student

In this chapter and the two that follow, I move to an exploration of the nature and range of learner identities among undergraduate mathematics students, an interesting group because, unlike the students who are the focus of Chapters 2 and 3, they have chosen to study mathematics, and chosen to do so within a high-status and high-stakes context. Their choice of degree-level study indicates a belief that they are at least at some level "good at maths," but what are the available identities within this group? What discourses do they draw on, and what makes for "success" in such a group? Can we trace individual trajectories in undergraduate students' positionings within their particular communities of practice? What is the role of the institution and the teacher in these positionings now that learning takes place in the university, not school? Are there connections to be made between their experiences and those of the students in the last chapter?

In order to answer these questions, I use data drawn from the focus groups and interviews with undergraduate mathematics students to explore the communities of practices that they are learning in, and the discourses of mathematics and learning mathematics that are at play within these practices. In this chapter, I explore the first-year undergraduates' stories, which show that a perception of oneself as a "legitimate peripheral participant"—as a novice with the potential to make constructive connections in mathematics—is difficult. Instead, these students tend to describe themselves as marginalized: they are aligned with mathematical procedures but do not contribute to them. I will argue that a positive learner identity is not necessarily associated with that of novice/apprentice as might be predicted by a community of practice model. On the contrary, students who describe identities of heavy alignment can appear unworried by their lack of participation in mathematics, successful as they are in the more dominant local communities of practice which are shaped by particular, familiar, discourses of "being good at maths" which equate ability and speed. It is possible to make sense of this puzzle by considering how individual trajectories through Discourse space and institutionally supported beliefs about ability and ownership of knowledge interact in the development of particular student identities, and the presence of gender differences in their experiences. I will suggest that gender discourses grow in positioning power as time goes on, and that issues of power and authority emerge as ones that must be addressed.

Experiencing Undergraduate Mathematics

In the post-compulsory years, identity persists as an issue despite the choice element in studying mathematics beyond the age of 16. As we know, many students do not choose to study mathematics despite good performances during the compulsory years, and research on this issue suggests some important features of their previous experience which undergraduate students bring with them to university and which may well have an impact on their experience there. For example, Steward and Nardi (2002, p. 7) argue that labeling pupils as good or bad at mathematics via the increasing use of standardized tests at an early age damages the likelihood of take-up when the subject is no longer compulsory, sending a message that "mathematics is hard." Gender is an issue, as we have already seen in Chapter 2: Landau (1994) observes girls' lack of confidence and the negative effects of accelerated GCSE courses on their subsequent take-up of mathematics; Kitchen's statistical survey (1999) notes that gender is a major factor in the changing patterns of A-level mathematics entry, with under-representation of girls who have good GCSE results at this level and also at undergraduate level; Mendick (2005b, 2006) also argues that "doing mathematics is doing masculinity"—for girls, choosing to study beyond the compulsory years therefore involves considerable "identity work."

When it comes to entering into university mathematics, the development of learner identities reaches a new level of complexity. I will review here the findings of two major studies of students' experiences of undergraduate mathematics, both of which examine the resources that students draw on in order to succeed at undergraduate level. Although very different—one is British, the other undertaken in the USA—they expose particular issues for women which act as a backdrop to my own data in this and the next two chapters.

The Student Experiences of University Mathematics (SEUM) project (see Brown & Macrae, 2005), tracked undergraduate mathematics students for three years in two British universities, focusing on their changing attitudes towards mathematics, issues in retention, student adaptation to university life and the role of social, economic and educational background in staying the course and final outcomes. The data from this project have been analyzed with respect to various issues: the characteristics of successful students (Brown & Rodd, 2004; Macrae & Maguire, 2002), and of failing students (Macrae, Brown, Bartholomew, & Rodd, 2003b); transition from school to university (Macrae, Brown, & Rodd, 2001); the experiences of female students (Bartholomew & Rodd, 2003; Rodd & Bartholomew, 2006); and the role of emotion and imagination in learning (Rodd, 2002, 2003). What unites these analyses are a number of themes which revolve around students' responses to university teaching and the resources they draw on in order to cope with its challenges; these themes find parallels in the US research which I describe below, and in my own data.

The transfer from the small classes and close teacher–pupil contact of the majority of British schools to the much larger scale, impersonal and lecture-based teaching style of university requires considerable adaptation on the part of students. Brown and Macrae (2005) report that those who were able to make the adjustment without reporting major problems were likely to be extremely enthusiastic about the intrinsic pleasures of the course, or, alternatively, to take a more instrumental view of gaining a good mathematics degree as a passport to a good job, focusing on gaining the best marks. Yet another group of students simply operated on the basis of doing well enough to pass at a level which they considered appropriate. For those students who were less than happy in their degree programs, a common factor was a reported decline in enjoyment. One cause of this appears to be the fact that many students choose to study mathematics at university simply because they find it easy at A-level, and so risk losing motivation when the work becomes more difficult and success is no longer guaranteed. That failing is more a product of difficulties in adjustment rather than a deficit in prior achievement and/or learning is suggested by the fact that Macrae, Brown, Bartholomew, and Rodd (2003a) were unable to find any obvious difference between failing and successful second-year students in terms of their academic qualifications at point of entry to university. Analyzing the same group of students, this time including interview data, Macrae et al. (2003b, p. 60) identified a "lack of academic preparedness" in terms of the motivation and study skills needed to withstand what this group of students described as a degree program which was "boring" and "hard work."

Given these demands, material and emotional support, together with a general understanding of university systems, is important. Successful students were more likely to come from graduate families and/or to be financially independent (Macrae & Maguire, 2002; Macrae et al., 2001), their cultural capital aiding not only their choice of university but also their transition to the range of social and pedagogic challenges that university presents. Macrae and Maguire (2002) also found that emotional support from family was an important factor in students' adjustment to university, and it seems plausible that this may offset other difficulties and increase resilience to the potential isolation and the pace of university study. Despite their experience of support from school teachers, students did not necessarily seek help from their university tutors because they expected tutors to approach them (and were in some cases disappointed that tutors seemed "not to care"). A minority of successful (male) students approached tutors confidently, but more in the spirit of dialogue than help (Brown & Rodd, 2004; Macrae et al., 2001). Thus the degree to which students draw on peers for tutoring and support is crucial: Brown and Macrae (2005) report that students who had more positive attitudes to studying mathematics were those who shared their ideas and problems with other students.

Students correspondingly appreciated project tasks which not only enabled collaborative work but also control over the pace of work and time for consolidation. Their liking for working with peers was exploited to good end in one of the participating institutions, where students were required to work on a group project in their first year. Some students also carried out individual projects in the third year and described this as useful and enjoyable because they could work at their own pace and in their own ways, reflecting that knowledge gained from project work was more likely to be retained than that learned for examinations. An important finding with respect to the data that I report in this chapter is that feeling part of a mathematical community emerged as a crucial factor in the student experience. In the SEUM project, this community focused on a particular physical space within one of the participating universities, a feature of undergraduate mathematics that I return to in Chapter 6.

Although the SEUM project did not set out to explore differences in the experiences of men and women students, these did emerge from the data. Bartholomew and Rodd (2003) report on the invisibility of women: they were less vocal but also less noticed (by tutors and peers alike) in the classes that they observed. In interview, these women found it difficult to identify as good at mathematics. Bartholomew and Rodd suggest that this invisibility is a result of the lack of a discursive space for women who do mathematics—the available identities are masculine:

> . . . the "effortless achievers" (high attainers who presented themselves as being unconcerned about their work, and not working particularly hard) and the "boffins" (who seemed to live and breathe mathematics, and made a point of making themselves known to lecturers and other students) . . . [But] What would a female mathematics boffin look like? How would her behaviour be understood by others? (p. 10)

Their analysis leads them to conclude that young women can only position themselves as good at mathematics by making themselves highly visible and stepping out of the available female identities: contributing in class, for example, triggers a particular response for one young woman: "on one occasion when she offered a simplification there was an audible 'oooh' from the class, suggesting she was being unattractively clever" (p. 17). In a related paper, Rodd and Bartholomew (2006) explore how young women in this contradictory situation have to "decide how to belong" (Griffiths, 1995), choosing invisibility as an intentional (as opposed to purely imposed) state which enables self-protection from the difficulties of "being a mathematical girl."

As these analyses suggest, students of mathematics will take up a variety of positions in developing their mathematical identities, and it is likely that successful students will not necessarily display identities of participation which neatly match Wenger's (1998) engagement model. Indeed, Brown and Rodd

(2004, p. 104) report a number of ways of participating in mathematics among their group of first class students, "their patterns of engagement being very different and their motivations varying hugely": some students in their sample focused on individual pursuit of right answers and instrumental application, while others relished mathematical debate. Their images of mathematics varied correspondingly, as "a meaningless game which is fun to do, maths as a source of the processes of following through tedious details, maths as a practical subject / a beautiful subject, or even, considered on a meta-level, as a high status subject that is character and mind-developing" (p. 104). It is possible, then, for highly successful students to display characteristics which are more closely indicative of learners on the margins of a practice, not learners on an inward trajectory towards engagement, or novices who are "legitimate peripheral participants," to use Lave and Wenger's (1991) terminology. What Brown and Rodd add to the mix is the recognition of the existence of a further practice in these students' lives that intersects with that of the community of mathematicians: that of the undergraduate community of practice. Thus they observe that their successful students can also be placed in relation to this community—some on the periphery, some more central.

These ideas are expanded on by Seymour and Hewitt's (1997) study of seven U.S. four-year colleges and universities, focusing on student experience in science, mathematics, and engineering (SME) programs, and the difference between students who stayed on these programs, and those who switched out of them. The starting point for Seymour and Hewitt's research is the under-representation of women and students of color in SME subjects, and the fact that these students are more likely to switch from SME subjects; I concentrate here on the issues for women, returning to discuss issues for minority ethnic students in Chapter 7. What is most illuminating about their analysis is that the problems reported by switchers were also reported by non-switchers—the crucial difference between the two groups was whether or not they had sufficient resources to stay the course. For young women, these resources were of a particularly subtle kind, involving the nature of their responses to the kind of positions which are available to undergraduate SME students. Their accounts of their experiences concern issues of belonging, identity and femininity which also appear in the SEUM data.

Switchers and non-switchers of both sexes cited as problems a lack or loss of interest; belief that non-SME courses are more interesting or better educationally; poor teaching; and the overwhelming pace and load of curriculum demands. These were the four most common contributors to switching, but they were also problems for between 31% and 74% of non-switchers (p. 32). In addition, three further issues were given as concerns by over one-third of both groups—choosing SME for the wrong reason; inadequate academic and pastoral support; and insufficient high school preparation. It is noteworthy that this latter problem was cited by 15% of switchers as directly contributing to switching, and by 40% of switchers as a concern (which may or may not have contributed to switching, therefore) and by 38% of non-switchers as a concern. Conceptual difficulties follow a similar pattern, with 27% of switchers and 25% of non-switchers citing it as a concern. Together with other data demonstrating the persistence and high SATs and college scores of switchers, Seymour and Hewitt argue that there is no evidence that switching is a result of "hardness" of SME subjects. Of the four issues which *did* distinguish switchers from non-switchers, three concerned career: reward is not worth the effort; SME lifestyles are not seen as attractive; and non-SME careers are more so. The fourth related to loss of confidence because of low grades in the early years (due to "weeding-out" practices). Thus it appears that there is little difference between the two groups in terms of the problems that they experience, and that these problems are overwhelmingly to do with the experience of teaching and learning at this level.

As I have already noted, switching is not related to "hardness," nor do switchers have lower entry scores. Focusing on women, there is no evidence that they are less competent than men, and in fact Seymour and Hewitt report that they tend to enter college/university with higher grades than men. Analysis of their reasons for switching show some interesting trends in that women are less likely to leave because of curriculum overload or pace (29% versus 42% in male switchers), because of low grades (19% versus 27% in men), or because of lack of peer support (8% versus 16% for men). However, they are more likely to leave because another subject is more interesting (46% versus 35%) or because they reject the lifestyle of their SME major (38% versus 20%), or because they feel that their teaching has been inadequate in advice or help (29% versus 20%). Seymour and Hewitt conclude that the figures

> Offer a picture of women in S.M.E. majors as students with lower instrumentality than their male peers, and with greater expressed concern about the quality of their education and their working lives beyond college. (p. 237)

Indeed, they go on to suggest that while both women and men complain about the quality of their teaching, they "diverge not in the perception that pedagogical problems exists, but in their definitions of 'good teaching', in what they expect of the faculty–student relationship and in the consequences of their unmet expectations" (p. 239). These are crucial differences, and they underpin other, often more public, differences which have far-reaching consequences in terms of women's self-positioning and the ways in which they are positioned by other students within the undergraduate community and the cultural models that it draws on. For example, women are more likely to seek out peer group help and to admit that they need help—Seymour and Hewitt suggest that this is one reason why they are less likely to leave because of low grades—they have "a buffer against the negative impact of the weed-out experience" (p. 238). However, this may have unforeseen consequences: Seymour and Hewitt also found that young men were loathe to admit to having to worked hard or that they found their studies difficult, and that perceived markers of inherent ability—apparently making little effort, not asking questions in class, avoiding peer study—were crucial in maintaining a position in the male hierarchy. Thus women's tendency to ask questions and to admit to problems consistently breaks the "ground rules":

> Women could unwittingly break the rules of the male status system by openly discussing their problems, or by asking questions in class. They were largely unaware that this reduced their claims to "smartness" among the men. (p. 251)

This unawareness extends to the possibility that young women do not receive the same messages from weed-out as young men. A major element of traditional SME is, Seymour and Hewitt suggest, the concept of "challenge," and the belief that students have to show their mettle before they can earn a place in a teacher–student relationship based on apprenticeship. They argue that young women do not know how to respond in accordance with the norms of this community: a central aspect of the challenge and weed-out ethos is competition, but, while it is apparently a motivator for young men, "what motivates most young women is neither the desire to win, nor the fear of failure in a competition with men, but the desire to receive praise" (p. 265). With the exception of black women, Seymour and Hewitt found that SME women sought the approval of others, and wanted to foster positive relationships with their teachers. Their accounts of what makes a good teacher were colored by this desire, stressing more than men the ideal teacher as approachable and interested in them as a person (p. 267). While young men understand that "the denial of nurturing by adult, male faculty is a temporary hardship" (p. 273), women do not. They have to learn not to care, as these two women mathematics non-switchers explain:

> Some of my girlfriends and I used to take it really hard when we didn't seem to do so well, you know, hiring tutors and just struggling and crying over grades . . . And it was all because, as hard as we tried, we just couldn't seem to please the professors. . . . Eventually, I learned not to take it to heart. It's not *you* they're grading: it's just your work . . .

> You'd look around the room before the exam, and the girls would be all silent and sweating. And the guys . . . most of them would be talking about football, or where they were going that night. They understood these were just weed-out classes, and they weren't looking for a pat on the back. (p. 269)

Coupled with the radical difference between their pre-university classes in which there was a balance between the sexes and the undergraduate classes in which they are in the minority, these differences in experiences and their response to them work together to make being a female SME student hard "identity work." Even when they did well, or well enough, many of Seymour and Hewitt's young women expressed feelings that mathematics was "no place for women," reporting that they had to deal with the stigma of being a girl who was good at mathematics, or a feeling of "not belonging":

> Women were also concerned that male acceptance of their academic worth would have negative consequences for their sense of who they are as women. The problems of belonging and identity are linked, because the qualities that women feel they must demonstrate in order to win recognition for their "right" to belong (especially "smartness", assertiveness and competitiveness), raise the anxiety that such recognition can only be won at the expense of "femininity". (p. 243)

More generally, women experienced, or were the object of, multiple tensions within the discourses of being good at SME subjects and being female. Rising to challenge by doing well is risky—it contradicts stereotypes and incurs negativity: thus Seymour and Hewitt argue that "female S.M.E. students who are conspicuously successful literally 'cannot win without losing'" (p. 262). Like Bartholomew and Rodd's SEUM women, many coped with this by keeping quiet about good grades and by making themselves physically "invisible." Another way of coping is to resist competition altogether: while some young women did engage in competition, many others refused to do so, seeing it as pedagogically counter-productive. However, survival overall appears to be a question of refiguring teacher–student relationships, learning to need them less, or finding support from others. This engineering switcher explains her own development:

> One reason I did well in high school is because I cared about what the teachers thought about me. I know I was doing well when people were pleased with me. I was always looking for that praise just so I know I was going okay. It took me a long time to get over that when I came to college. I used to get very upset because, here, the teacher doesn't know who you are. . . . Now I love just knowing that I am doing well—but that's not how I started out. (pp. 260–70)

And this engineering non-switcher explains the value of gaining support from other women, often through collaborative work:

> The structure of the courses are very geared in a very male way—with individual problem sets. . . . It's all kind of directed towards doing it by yourself . . . Still, the women work together a lot. We always worked the problem sets together in small groups. (p. 299)

Seymour and Hewitt's research and the SEUM study together indicate that undergraduate mathematics identities need to be understood in terms of the interface between different practices and different discourses, some of them diametrically opposed or contradictory, and in terms of the ways in which different practices make different identities available. In the analysis which follows in this and the next two chapters, I explore the complexities of the communities of practice which undergraduate mathematics students are party to, or potentially are party to, via their different modes of belonging as the communities intersect. First, I examine the mathematical identities of a particular group of students in transition: first-year undergraduates are an especially interesting group because they are new to the university community, bringing with them their immediate mathematical histories. In addition, the particular group studied here are in the position of being able to make further choices of what to study next, and so there is an extra dimension to be taken into account in terms of how their experiences and self-positionings relate to their proposed action. A close look at their stories presents an opportunity to answer some of the questions raised at the beginning of this chapter. What emerges from the analysis is the presence of dominant discourses which underpin the values and beliefs of the undergraduate community and constrain the available range of identities for this group of students.

Undergraduate Mathematics Identities

In this section I explore the learner identities of the first-year undergraduate mathematics students with respect to the communities of practice within which they function, comparing their accounts in terms of Wenger's (1998) three modes of belonging—alignment, imagination and engagement—and combinations of these. Exploring student identities in this way emphasizes two important aspects of mathematics learning. Firstly, it makes transparent the role of beliefs about mathematics and mathematical abilities in the development of identity. Secondly, gender differences emerge which suggest that classroom communities and practices have a considerable effect on the development of identities of alignment, imagination and engagement, and how these are experienced. I will suggest that what makes a functional identity in this particular group of students—that is, a perception of self as able to succeed in undergraduate mathematics—is not necessarily an identity of potential engagement, or, in Lave and Wenger's (1991) terms, legitimate peripheral participation in the wider world of mathematics. Paradoxically, within this community those students who do aim for a more participatory role may in fact doubt their ability to continue as mathematics undergraduates, developing identities of exclusion, rather than inclusion.

Following Rules—Negative Alignment

Alignment to a practice emphasizes common agreed systems of rules, values or standards through which we can communicate within a practice and through which we can belong to it. However, while alignment is in this sense positive, enabling us to function within a practice, systems which we do not own and cannot contribute to are no more than rule-bounded situations in which we participate only as rule-followers, not rule-makers. Although initial guidance and modeling introduces the learner to the possibilities of a practice, lack of ownership generates and is generated by compliance and an emphasis on procedural or "ritual knowledge" (Edwards & Mercer, 1987, p. 99) which is "embedded" in the context in which it has been learned and is, therefore non-transferable. As I noted in Chapter 1, Wenger (1998, p. 206) argues that while on one level this is efficient for the learner precisely because of their lack of ownership, its dependence on its original context makes it inflexible when new situations arise.

Indeed, a number of students described their mathematics activities as blind rule following, but they varied in terms of whether they experienced this as a source of irritation or were accepting of the situation. Such variation appeared to depend on other aspects of their identities, generated from their views of their own abilities and dispositions, and from their classroom experiences. For instance, Steve considers rule-following unproblematic, and even a bonus:

> I like learning methods and, like, getting just one answer . . . As a person I don't really like making decisions, I like everything laid out for me.

Charlie equally sees no problem in rule-following without the support of intermediate steps:

> If I've got the knowledge . . . it's—like—learn and just memorize it. I hate the long way of doing something and then there's an easier way, I say you're never going to use it again so why did they teach it you in the first place?

Although he values understanding when it underpins being able to do the work, Joe has a similar attitude to Charlie with regard to tutor demonstrations which are not assessed, and he describes how he is not interested in how results are arrived at if he will not be required to reproduce these steps in the examination. Chris too is not bothered about understanding as long as there are rules to follow:

> [If] I'm told so and so and so and so is this, then I won't go and read and try and understand why, I just remember the result . . . I just accept what people say . . .

As these extracts indicate, it is possible to get right answers through mechanical means, and Chris in particular was satisfied with this, relegating understanding to a lesser priority. Richard, who was very focused on his test results, liked mathematics because he could do it and get it right. He has a clear preference for a subject which he perceives as having right and wrong answers which are given by the rules, and are outside the realm of opinion or debate:

> I don't care as long as I can do it. . . . What I like about it [is] the fact that it gives you a right answer. If there's a definite answer, I'll be alright. . . .

While he continues to get right answers, Richard is unconcerned; however, his tolerance for failure to get the right answers is minimal:

> I was alright at English but as soon as my grades started going down I dropped it. I just don't enjoy doing something I'm not good at.

Larry tells a slightly different story although his history of mathematics clearly hinges as Richard's does on his ability to do it. He says that he does not particularly like mathematics—like many students, he does it because he can:

> I wouldn't be too sure that I like maths. I guess I carried on choosing it because it's been easier. I do find some of it interesting but I couldn't say I like it enough to get a career in it.

In contrast to Richard, however, he seems unworried about his marks:

> It varies. I've got above 35%. I'm not too fussed about marks in the first year—if I get above 35% it means I can cope with the second year.

A very different feeling about rules and achievement is described by Sue, however, who experiences mathematics as confusing and pointless. Her irritation and frustration with the situation and her perceived role in it is situated in her experience of her mathematics classes, where, for her, the need to follow rules without understanding signifies marginalization from the wider community of mathematics rather than belonging:

> . . . you have to do hundreds and hundreds and hundreds of examples and they get harder and harder and harder and you end up confused—things like natural log and integrating and all sorts of things—but there doesn't seem to be any sort of reason . . . and every time you think you're there, they say "Oh, this is just an introduction, this is what you really need to know how to do", and I think, "Oh, this gets worse every time". . . . Analysis—integration, double integration, again, they don't have a point,

I don't see the point in them at all, which makes it worse, you know if you can't do something, you don't like it then do you, so I don't like them because I can't do them, and because they don't have a point I don't like them even more, they come up everywhere.

While Sue maintains a hope that she will understand more if she persists with mathematics and tries to become party to the meaning behind the rules, Carol's similar frustrations lead her to resist and reject it. Her experience of non-participation in class results in her description of learning mathematics in terms of "them" and "us," particularly when it comes to pure mathematics, which, she believes, does not allow her to express her own understanding:

I always wonder about maths, because I'm not really the kind of person that just accepts things, I always like to see the proof of it all and they just reel off all this stuff—"And this is how you do this"—and I'm, "Well, why?" . . . Calculus: different styles of integration—do they explain why? No—"they just are". Which is useful (sarcastically) . . . Probability and stats you can do more hands on, you can do more work yourself, you can have your own data you can do your own thing.

Reflecting on Mathematics: Imagination and Engagement

A few students expressed a more positive relationship with mathematics, reflecting on their position within the wider practice of mathematics, on the patterns within it and their identity as learner mathematicians. They thus showed what Wenger (1998, p. 176) calls imagination: a positioning of self within the social nexus of practices and an awareness of actions as part of historical patterns and potential future developments, of others' perspectives and of other possible meanings—the self is expanded by this reflexivity. They willingly reflected on what they were doing as students of mathematics, and considered explicitly how their approach helped them, often with reference to their own particular learning styles and experiences. Larry, for example, reflected on his success in mathematics:

It felt like you could condense maths to quite a few methods/facts which would be useful; for solving things whereas chemistry you'd be using [a wider range of things]. Maths is based on a set of facts which you can follow through without having to rely on other things, you could use past experience.

Here, Larry expresses an identity of being in control of learning mathematics. Debbie, while not understanding as well as Larry appears to do, enjoys the exploration of mathematics. Here she describes her feeling about mathematics in general and her discovery of proof:

[With mathematics], I feel like I can get an answer to something . . . And it's nice to, I get sort of relief from it actually. It sounds a bit strange but, you know, I enjoy to work something out and get an answer at the end of it, it gives me great pleasure . . . [Proof], that's another thing where . . . it's giving a reason for something, I like to have a reason for it and that's something very new to me, I'd never come across it before I came here. . . . I didn't even knew it existed actually [laughs] before I came here, didn't know there was proof . . . so, yeah, it's sort of a new thing for me.

A true sense of engagement in terms of an attempt to "appropriate the meanings of a community and develop an identity of participation" (Wenger, 1998, p. 202) is evident in Sarah's account only. While she describes a preference for the security of being told what to do by tutors, she also finds herself independently looking for patterns and exploring them:

I find it a lot easier, you know, for them to say "this is what you are going to do, and this is how it's done" so in a way I am not very creative in my maths, if that makes sense, but in a way I am as well because I actually, sometimes I'm working and I think "oh maybe this could work" you know, this and that, and I get all excited and it usually doesn't work but still I am thinking about it.

Accounts such as Sarah's are rare. Notice the contrast between her willingness to be guided by her tutors and Carol's far more adversarial stance; while both these students seek some sort of self-expression in mathematics, Carol seems to perceive herself as excluded, as wasting her time, but Sarah sees herself as potentially included, and indeed acts as though she is included. Her attitude to mathematics is correspondingly different; she appreciates its aesthetic:

It is nice. And also at the end you have this nice thing and you have worked all through it.

Positioning Identity within Multiple Communities of Practice

In the preceding analysis I have selected comments which illustrate particular modes of belonging, most frequently a negative form of alignment, but also modes of imagination and engagement. As we have seen, the students tended to describe themselves as lacking control over their mathematical knowledge, as following rules without understanding, and as vulnerable to failure—staying with the subject is possible only as long as they can do it, and this facility can fail at any time. It is in this sense that most of the students express identities of marginalization in an alignment to mathematics procedures which they learn to operate but do not contribute to. Only one student—Sarah—described herself in terms which fit the label of a "legitimate peripheral participant" who, as a novice, has much to learn but also has the potential to make constructive connections in mathematics and to act as a negotiator in the mathematics community. However, there is a distinction to be drawn between membership of the wider community of the discipline and of the various other communities of practice which an undergraduate student is likely to come into contact with. The characterization of student identities above holds only with respect to the community of professional mathematicians of which some are only dimly aware and/or may not aspire to be a part. There are more immediate communities of practice which also figure in these students' identities: the undergraduate community in general, the mathematics undergraduate community and the first-year community within it, and the classroom community of learners and tutors. The students' identities and their relationships to mathematics are also shaped by their membership of these often more visible communities.

The analysis thus far also shows that their modes of belonging are not experienced in the same way by all students: we need to capture an individual's position with respect to multiple communities of practice in order to fully understand the complexities of mathematical identities. While Wenger's model attempts to capture complexity in its definition of identity as "a layering of events of participation and reification by which our experience and its social interpretation inform each other" (Wenger, 1998, p. 151), it neglects to explore in detail the nature of identity in multiple, and possibly conflicting, communities of practice which share the same "Discourse space." Building on Gee's idea (2001, p. 111) that we each move through a unique trajectory through Discourse space and that we build our narratives of self on our unique experiences within specific Discourses, we can however make some sense of apparent conflicts in what the students say.

While my discussion of Sue above illustrates an identity of not belonging in terms of her negative account of rule following, a more holistic reading of her interview shows a clear element of imagination: she reflects on her experience of learning mathematics, and attempts to make sense of it as part of her university experience, and to situate it in her own mathematics history and future. There is a mismatch between the values of the wider community of practice of mathematics and those of the immediate undergraduate and classroom communities of practice which Sue is part of. Although she described herself throughout her interview as confused about mathematics, her attempts to understand both mathematics as a discipline and the merits and demerits of her particular approach to it suggest perhaps that she is not so much marginalized from the wider mathematics community than from the undergraduate and classroom communities, with their emphasis on getting right answers, following rules, and speed. Here she reflects on the difficulty of undergraduate mathematics, attempting to draw on resources, knowledge and experiences that she has developed elsewhere:

> With physics and chemistry moving from GCSE to A-level, the things we accepted at GCSE were then explained at A-level . . . we still used them at

GCSE but we just had to accept that . . . you need to know this now but you can't understand them yet. . . . In maths . . . maybe it's because I haven't studied the whole picture, I've just got this little bit, and with being told I've got to accept things, and I've just accepted things as I've gone along, and now I've got to a point where I can't just accept things. I need to understand things that I'm being told, but I've got to accept a little bit more before I can start understanding . . . Maybe if I carry on doing maths it might click again.

Unlike Carol, Sue accepts the apparent inconsistency or opacity of mathematics rules, assuming that they do make sense if only she can stand back and take in the wider picture. As yet unbeaten by the challenge of confusion, she displays imagination in the sense that she is able to "accept non-participation as an adventure" (Wenger, 1998, p. 185). In the remainder of this chapter, I will explore how individual student identities come to display a particular mix of modes of belonging by examining the interplay of structures, practices and cultures. We can then begin to understand what constitutes a functional learner identity, in the sense of belief in oneself as a learner, in the undergraduate mathematics community of practice.

Fixed Ability Beliefs

While Sue's account shows how a negative alignment experience can be offset by imagination, she nevertheless hovers on the threshold of a negative mathematics identity, a characteristic she shares with others in the group. In large part this fragility is due to the almost universal fixed ability beliefs which are perpetuated by the pedagogic practices that surround them and permeate the undergraduate community of practice. Thus Carol, in spite of her robust and critical outlook on mathematics teaching, puts her non-participation down to perceived deficiencies in herself; her beliefs about ability and the nature of mathematics itself all militate to build a self-excluding identity:

I don't know whether I've got to the stage where I think it's too difficult or I'm not bothered any more or if I don't really see the point of doing it any more. . . . I think with maths, you're good at it or you're not particularly good at it . . . you can struggle for years and years to understand maths and never grasp the concept, I think it is an all or nothing subject.

Ultimately, Carol subscribes to the idea that "you can either do it or you can't." Despite her rather more sophisticated and imaginative recognition that mathematics is about making connections, Diane also believes that successful students have a built-in overview of mathematics which enables them to solve novel problems:

I think that they can bring all the bits of maths that they've already done together whereas I think I need someone to say "You have to take this from here and this from here and put them together to work out the answer to this one" . . . I think people who are good at maths can recognize that already and use the information that they've already got . . .

She invokes brain functions in mathematics ability when asked whether her mathematics performance could improve:

I could be taught to do maths but I don't think I could be taught to be good at maths. I think that's just something about the way the brain works or something.

Pete also believes in a biological basis for being good at mathematics, claiming that good students "have some innate ability." Sarah stands out from the other students in her claim that hard work can reap benefits in terms of developing a mathematical way of seeing:

Do you think that you can improve as a mathematician? Yes, I think I could . . . I definitely think I could put more effort in and . . . go through and look at all the different examples and what happened in those examples and by doing that you learn and you learn to be able to see what is going to happen.

Nevertheless Sarah still invokes the idea of an uncoached mathematical "talent" which echoes Diane's, and she contradicts her claims about hard work by giving voice to another dominant belief in the undergraduate community, that really good mathematicians never fail and don't even have to try:

I think some people can and some people can't . . . [They] usually don't do much work at all . . . they leave it to the last minute and they just do it and then they get full marks . . . I am good at maths compared to most people but compared to them I am awful because . . . they just have the mind for it, they can just see.

Debbie also refers to the companion belief that good students are fast workers:

You know, some people . . . you get the impression that they don't really even have to think much. . . . I don't think that my brain is as clued up as some of these that obviously can just do it.

The prevalence of such beliefs in these students' accounts alongside indications of imagination and engagement suggests a reason why students such as

Debbie, Diane, Carol, and Sue struggle to maintain a positive mathematics identity despite their apparently more participatory trajectory into the wider mathematics community. While Diane, for example, recognizes explicitly the need to make connections in mathematics, the discourse of fixed ability, performance and speed of understanding which pervades this undergraduate community has a detrimental influence on her identity. Sue interprets her need to understand as problematic, while Carol believes that she has reached her limit. Even Sarah describes herself as "awful." Looking closer at what these students say about the institutional structures and practices that they are part of shows how these continue to support the notion of fixed ability, thus undermining potential participation in mathematics and creating identity mixes which are *experienced as* marginalized. I explore these issues further in the next section.

Institutional Structures and Practices

Identities of non-participation in mathematics have important consequences. As the first-year accounts illustrate, these students experience mathematics as something "done to them" rather than "done by them"; they do not share in the ownership of meaning, let alone meaning making—they are excluded from that vital aspect of participation which Wenger identifies: negotiation. Engagement in a practice entails an identity which includes the role of legitimate negotiator of meaning—those who participate fully in a practice are part of the process of development of ideas and meanings, and in this sense have ownership of meaning. To some extent this is also the case for legitimate peripheral participants—their ideas and contributions are treated as valid, to be taken seriously, to be built upon. However, the majority of the students do not perceive themselves as potential negotiators or owners of meaning. It might be argued that undergraduates cannot expect to find themselves in this position anyway, but many critical mathematics educators have argued that this is not only possible but necessary from the early years onwards (see Maher, 2005 and Burton, 1999a) to HE levels (see Rogers, 1995 and Schoenfeld, 1994). To take an example from this body of research, Schoenfeld (1994, p. 66) demonstrates that a plausible goal of undergraduate mathematics classes is to "reflect (some of) the values of the mathematical community at large . . . [and to] create *local intellectual communities* with those same values and perspectives . . . a contribution is significant if it helps the particular intellectual community advance its understanding in important ways" (my emphasis). This is an important distinction—clearly, undergraduate students are unlikely to be able to contribute to the mathematics research community, but this does not have to mean that they cannot participate in their own community in parallel ways.

The important issue, then, is how undergraduates experience and make sense of their classes, and how this influences their self-perception and choices. A number of them reported that mathematics, particularly pure mathematics, was presented as a non-negotiable finished product, as a set of rules and strategies to be learned, not constructed. The net result of this teaching strategy was that pure mathematics was generally perceived as "hard" and—more importantly—as a subject which they could not contribute to or be creative in, or even simply catch up in. For example, Joe complained that:

> The lecturers are always setting us more challenging work which we don't understand . . . you never really feel like a mathematician because you don't understand how it works.

Sue expressed bewilderment as she described getting answers but not owning the knowledge:

> [It's] very frustrating, because you know you know how to do it, it's just the problems are so much more complex and they sort of go in more, I don't know, just things from nowhere, and you do get the answer in the end but you just don't know.

Diane similarly reports confusion and isolation as she compares herself to the "good" students who, as we have already seen, are fast workers:

> They seem to know exactly what to do and they're just integrating and differentiating all over the place and I have to wait for the lecturer to do it. That's why I think I'm not good at maths. . . . There's this one guy and I'm sure he's a mathematical genius, I'm sure he works it out. But I just wouldn't.

Carol and Diane both compared the mathematics department teaching adversely with that in environmental science and geography:

> You can't feel like a mathematician until you've learned quite a lot of stuff. [In environmental science] you're asked what you think about things. (Carol)

> In geography they just want to see that you've understood the question and see if you can bring your opinion into it. Whereas maths it's to see if you've understood. Full stop. (Diane)

Why do students such as Richard seem to be happy with their apparent alignment, and why is Sarah's evident engagement tempered with an identity of marginalization? The gender differences in response to mathematics pedagogy observed by Seymour and Hewitt in particular seem to be at play within the various communities of practice of this first-year group and their classes. It appears that Richard, Steve, Joe, Chris, Pete, and Charlie experience positively the atmosphere of reward for speed and correct answers, and in this sense they are full participants in the undergraduate community of practice and its related pedagogy—they feel at home with mathematics as it is taught in this university in large, anonymous groups. This group comprises both intending majors and minor students—we might expect that intending majors would feel differently but this is not the case—indeed Larry is the only male student to state that he is bothered by not understanding. As Seymour and Hewitt found, women were more likely to demand understanding and engagement, and were dissatisfied by the intense competition which they experienced instead. The women in the group emphasized how much they wanted to understand, and their accounts were dominated by a sense of constant danger of feeling out of their depth—ownership is important to them but always threatens to be unattainable. The transfer from school is, then, experienced as a loss of space for understanding and collaborative endeavor. Thus Diane wistfully remembers her school days of small group support for understanding in an account which is reminiscent of Maher's (2005) work on proof with high school students:

> In my A-Level . . . we'd all work together to get the same answer and I think that really helped because we were teaching each other which would help us to understand. . . . Because each of us understood different parts of it we were like, "No, no, you're wrong, you're wrong", and say, "Well explain yourself then". . . . I think it really helped me get through A-Level because you learn from each other as well . . .

Her outlook contrasts starkly with Richard's:

> I think I'm the kind of person who should care about understanding but I don't . . . I am competitive . . . getting the right answer is more important . . . I understand well enough to carry on.

While the men tended not to raise issues of teaching or group dynamics, the women were very likely to volunteer their appreciation of the value of working in a group, partly because the group lends reassurance that they are not alone in not understanding. Carol described why she informally sought out other students when she was stuck, highlighting at the same time the lack of discussion in class:

> I think it's just reassurance that you're not completely stupid because you can't do it, and just bouncing ideas off another person is better than sitting in your room attempting a question 50 times because you don't know how to do it. . . . It's easier to talk amongst yourselves [outside of lessons] whereas in a tutorial you kind of feel under pressure just to not say anything in case it's the wrong answer.

Charlie admitted that the norm was for the group as a whole to avoid interaction in class because "You don't want to look stupid if you're doing something really simple." Diane in particular comments on the need to make and discuss links between ideas in mathematics, but her belief that she is unable to do this causes a major loss of confidence. Rather than supporting a participant identity, then, her insight into the role of making such connections causes her to feel increasingly inadequate and marginalized, not the reverse. Why should this be? It appears that the structure of mathematics learning as she is currently experiencing it disallows access to its meaning-making practices— there is no time to do so and the reward system is not geared to it. In terms of this analysis, the immediate undergraduate community of practice does not enable legitimate peripheral participation in the discipline of mathematics. Rather, it marginalizes learners who seek to participate beyond a focus on correct answers, causing them to doubt their ability. Diane, among the many comments in which she compares herself adversely with those who are "good at maths," says that the "good" ones are the confident ones, and they are usually male:

> [They are] usually men . . . they're getting too big headed and they know "I can do this" . . . They're all smug and they sit there and they're filling in the answers and then they sit back and sort of look over at what the other guy who's sitting next to them . . . like, "Huh, you've done it wrong there" . . . Some of them are just really confident that they can do it and then they do it and they're really good.

Similarly, Sarah comments that students who will speak publicly in lectures are much more likely to be men; if they notice a mistake, the women would not normally speak out, whereas the men will:

> I think they are more likely to be the ones that are going to point out there is a problem, you know, "there is a mistake on the [board]" or something like that, I have never seen a girl do that, well I have done a couple of times but I never really, I wouldn't do it in a lecture . . . they'd probably just leave it, or, you know, say to the person next to them, "that's wrong" or something like that but I wouldn't think they were going to shout it out unless they are quite a woman.

Excluding Practices and Identities of Exclusion

What generates identities of non-participation and/or marginalization? Much of the research that I have reviewed in earlier chapters indicates that school mathematics teaching is frequently excluding, and that it treats many students as powerless and unimportant "outsiders," apart from the few who earn a place in top set or further mathematics groups. We might suppose that students who choose to study mathematics at undergraduate level would experience it more positively, but the findings of the SEUM project and Seymour and Hewitt's research indicates that the situation is more complex than this: not only do students frequently criticize their teaching, finding it difficult and sometimes boring, but they also respond to this situation in crucially different ways. Despite their choice to study mathematics, an identity of legitimate peripheral participant is rare in this group of students. The analysis in terms of modes of belonging presented here shows how identities are differentially experienced within multiple communities of practice and hence goes some way to explaining why this is the case.

Identities of exclusion are most obviously voiced by the women in this group although they are less marginalized according to Wenger's model than the men in terms of their approach to the discipline of mathematics. They seek an engagement with mathematics which is epitomized in Diane's quest for links and patterns and Sarah's perception of the possibility of creativity and negotiation which many educators believe is necessary and yet is not supported in the undergraduate community of practice. This community draws on particular discourses of mathematics and learning which do not give them the means to develop their understanding, maintaining instead an explicit support for a performance orientation of the kind demonstrated most clearly in Steve's and Richard's heavily aligned accounts. What distinguishes the students' accounts is that, for the time being anyway, Richard, Charlie, Joe, Chris, and Steve—and to some extent Pete—are happy with this state of not belonging, and indeed do not express excluded identities, while Carol, Sue, and Diane are not. Even Sarah, the most confident of the women, believes that some people "just have the mind for it," and does not count herself among them. Richard, Charlie, Joe, Chris, and Steve accept their state of strong alignment whereas Carol, Sue, Debbie, Sarah, and Diane strive for imagination and engagement. To the extent that their experience of mathematics teaching and learning emphasizes speed and performance, the men in the study have the more functional identities in the undergraduate community of practice in terms of their belief in their ability to succeed in accordance with these values, and they are full participants in this community. The women, on the other hand, face failure regularly as they strive to meet the twin criteria of speed *and* understanding.

Thus the position of promising novice represented by Sarah's account is not associated with a positive identity, while that of the heavily aligned and non-participating Richard does not bring with it a negatively experienced identity. The resolution of these contradictory findings is brought about by recognizing their position in multiple communities of practices with opposing rules of engagement, and consequently of differential experiences of identity. As Brown and Rodd (2004) show, there is more than one way of being a successful student in undergraduate mathematics. The analysis presented here indicates, however, that some potentially successful students develop negative relationships with mathematics which marginalize them and can turn them against further study in ways which are similar to Seymour and Hewitt's switchers. Within the undergraduate community of practice, the dominant discourse of performance within which mathematics identities are constructed dictates the apparent functionality of particular identities. In the next chapter, I explore the interaction between beliefs, values and the undergraduate experience in more detail.

Doing Undergraduate Mathematics
Questions of Knowledge and Authority

In the previous chapter, I began to explore how students draw on particular discourses of mathematics as they construct and take up the various available spaces for mathematical identities at the undergraduate level. I continue this theme in this chapter, examining their epistemologies of mathematics in more detail, and their narratives of mathematics teaching and learning. A number of recurrent themes and connections emerge from the interviews with the first-year group which are indicative of particular patterns in their beliefs about the nature of mathematics and their own relationships to it as learners in the university mathematics community—both are encapsulated in what they say about proof, as I will show in this chapter. I will suggest that the students' self-positionings with regard to mathematics in general and proof in particular are best understood when viewed within the context of the discourses of power and authority which dominate their accounts, and which appear to foreground some epistemological frameworks while obscuring others.

The Undergraduate Community: What is Mathematics Anyway?

We have already seen that a strong and pervasive belief in the difficulty of mathematics and the related claim that "you can either do it or you can't" had major consequences for the ways in which students position themselves with respect to doing mathematics. Their more general beliefs about learning are equally embedded in the cultural models on which they draw, and connect to particular perceptions of the teacher–learner relationship.

Beliefs about Mathematics: Certain Knowledge, Learning Rules and Individualism

The undergraduate students' beliefs about mathematics are very similar to those that we have seen among the secondary school students in Chapter 3, and beliefs such as those observed by Schoenfeld (1992) and De Corte et al. (2002) are clearly discernible in what they have to say about their university experience. As the students describe doing and learning mathematics, a general theme of certainty in mathematics emerges, coupled with an emphasis on the necessity of learning rules, reproducing solutions and working at speed to get correct answers: a major focus of their learning life is doing "homeworks" and going over the solutions in the next week's session. Correspondingly, the students tended to claim that creativity in pure mathematics was not possible, although statistics afforded some opportunities in this respect. For example, as we have seen in the previous chapter, Richard was emphatic about the certainty of mathematics in contrast with his management course, demonstrating a strong claim to negotiable knowledge in management (of which he himself could be a part) versus a certainty outside both himself and his teacher in the case of mathematics: "There's a right and wrong in maths . . . there's nothing that's open to the teacher's opinion . . ." All of the first-year group were studying statistics as part of their course, and made similar comparisons within the mathematics curriculum itself. For example, Steve contrasted an assumed certainty in pure mathematics with a greater need for interpretation in statistics in which he ran the risk of "not being quite right." Pete also expressed a strong dislike of statistics because of its lack of precision, while Sarah described statistics as "a bit fiddly" in comparison to pure mathematics which is "do this, do that."

While these students disliked the need for interpretation in statistics, it was precisely this quality that the students who favored statistics over pure mathematics liked. Chris and Joe saw statistics as allowing for more autonomy and as being more meaningful. Thus Joe claimed that "there is a form of arguing in stats" in contrast to pure mathematics which for him was devoid of argument and dominated by rote-produced solutions. Carol favored statistics for the same reason, arguing that it offered more scope for creativity, "whereas we're not going to discover anything new to do with pure maths." Students who were pro-statistics also appreciated it as having potential applications to real-world problems, arguing that, conversely, pure mathematics had no applications at all. Both Carol and Charlie felt that this perceived aspect of pure mathematics detracted from any potential interest it might have. Sue also expressed a preference for statistics based on its more visible real-world applications in comparison to the difficulty and exclusivity of pure mathematics in which "there doesn't seem to be any sort of reason."

These views were shared generally among the group. Even those students who actively preferred pure mathematics to statistics concurred with the others in making a clear distinction between statistics as useful and pure mathematics as generally little more than a puzzle with intrinsic value. This general emphasis on pure mathematics as a rule-bound and largely useless activity in which they could not participate in any creative sense spilled over into images of mathematics as an isolated pursuit and mathematicians as rather cranky individuals, very different from themselves and fitting the stereotype that I noted in Chapter 1. For example, Carol described research mathematicians as "just sitting in their office with a calculator for 4 hours a week or something. . . . You do get a image of them in a room with a calculator for too long and that's why they're all going slightly insane . . . I respect anyone who has an interest in their subject but I think some of them are a little odd."

As these extracts illustrate, the students tended towards epistemologies of mathematics which assumed certainty, irrelevance, rule-boundedness and lack of creativity potential (for ordinary mortals anyway) in pure mathematics. For some this was tolerable, for others it was not. It may well be significant that the students who favored statistics on the grounds that it was not prone to these particular perceived characteristics then proceeded to reject mathematics as a whole, changing to other major subjects in their second year. It seems unlikely that this is merely an issue of subject preference—all the pro-statistics students had chosen to study mathematics at degree level, three of them originally intending to study it as part of their major options. An alternative explanation is that their university experience did not foster a sense of how they could become part of the mathematics community of practice, with the result that they opted out altogether rather than remain on the margins. Indeed, the students seemed largely unaware of—or could not imagine—the existence of a mathematics community which might have negotiable rules of communication and validation beyond the simple authority of the individual teacher-experts with whom they came into contact. I return to this issue below.

Beliefs about Learning Mathematics

In their account of beliefs about the self in relation to mathematics, De Corte et al. (2002, p. 307) differentiate between goal orientation (I want to understand), task value beliefs (learning the material is important), control beliefs (learning is possible with proper study), and self-efficacy beliefs (I can understand difficult material). While we might anticipate that making a free choice to study mathematics at university would be associated with strong and positive versions of these beliefs, this was not evident in the students' accounts. Nor were they necessarily intrinsically motivated to study mathematics: Charlie, Richard, Joe, Chris, and Larry all said that they had chosen mathematics at university for its value in the labor market. Sue, Debbie, and Sarah also made reference to the relevance of mathematics to their career plans. Richard presented a particularly interesting case because he had lacked the confidence to major in mathematics at university, choosing management science instead as a good career option, despite the fact that mathematics was his best subject at school. However, his first-year performance, in which he was consistently scoring higher than most other students, had persuaded him that he was "quite good" at mathematics and that he should change his intended major. This decision was very much based on comparison with others and an explicit prioritization of good marks over understanding:

Sometimes I don't really understand the process, but if I can apply it I'm happy with that. If I was struggling I'd drop it. . . . I have to be the best.

[*Doesn't that make you vulnerable?*] That could be the reason I didn't take it in the first place.

Comparison with the performance of others is a recurrent and explicit theme in Richard's account but it is also very evident in what the other students say. Pete invoked the familiar theme of speed of understanding as a defining characteristic of those who are good at mathematics, while Sarah illustrated its companion attribute of succeeding on the basis of very little work, claiming that the students who are good at mathematics can get top marks with no effort. Debbie queried this claim, but nevertheless suggested that perhaps such students had other personal qualities:

I think it is more of a belief. . . . I think that it's not cool to sort of say that you do any work before exams and stuff like that so I think with some of these people who say "oh, I didn't do any revision" . . . I don't know if it's really true . . . or maybe they've just got good memories.

The emphasis on speed of understanding and working is linked to the predominant fixed ability and "natural talent" beliefs which we have already seen in Chapter 4. For instance, Pete believed that "at university it's requiring more innate abilities" while Sarah thought that the good students "just have that skill . . . their mind is different or something." Given these views, it is not surprising that control beliefs and self-efficacy beliefs were not particularly evident. Diane, who was planning to drop mathematics in the following year, did not believe that any amount of effort would enable her to achieve the same as the "good students"; as we have seen in the previous chapter, she believes that while she can rote learn to do mathematics, she cannot learn to be good at it. Larry, who was intending to continue with his single major in mathematics, was a lone voice in describing a high level of self-efficacy when it comes to solving a difficult problem; he was also the only student to describe mathematics as an integrated subject rather than a set of isolated facts: "if you know a method—a method that makes sense—you can combine that to get something else."

A Case-study of Proof

As research in school mathematics has shown, learners can be, and often are, excluded from the negotiation of meaning or even the beginnings of it, developing instead an identity of non-participation and marginalization. Their lack of ownership generates and is generated by compliance with authority and an emphasis on following pre-set procedures which are reflected in the epistemologies of mathematics noted above. In making the move from school or college to university, students are in the position of entering into a novel community of practice in which success will depend on their ability to gain access to its ways of co-constructing knowledge—in the case of mathematics one of these central meaning-making practices is of course proof. It appears, however, that their previous educational experience has not equipped them very well for this move: the match or mismatch between their developing episte- mological beliefs about mathematics and those implied in degree-level and research-level mathematics will be of central importance, as will their more general beliefs about their own learning.

The ability to recognize, understand and deal with proof is frequently the focus of studies regarding students' difficulties in encountering degree-level mathematics (Almeida, 2000; Anderson, 1996; Cox, 2001; Kyle, 2002; Seldon & Seldon, 2003; Sowder & Harel, 2003). A wide range of research correspondingly details a lack of preparedness in school and pre-university students, and an emphasis on empirical rather than deductive proof schemes identified in the USA by the National Assessment of Educational Progress and other studies (see Harel & Sowder, 2007) and in the UK by Healy and Hoyles (1998, 2000). This latter work found that although proof was part of the statutory curriculum (recently updated—see QCA 2007a, 2007b), it appeared only within the context of investigations, which are data driven and involve informal argument generated from the observation of patterns. There tended to be no explicit or systematic engagement with proof and its usage. Recio and Godino's (2001) examination of the contrast between institutional and personal meanings of proof in Spain provides related evidence that students spontaneously and by preference use empirically-based proof schemes, particularly when presented with new or more complex problems.

However, in addition to observing that proof has a minor role in UK pre- university mathematics curricula with a possible skills deficit for the under- graduate mathematician, Almeida (2000) also makes a further important point regarding expert practice which refines the view: while, as MacLane (1994, p. 191) points out, the research mathematician handles proof procedures which include as a matter of course *intuition, trial, error, speculation, conjecture, proof,* undergraduate *teaching* stresses instead a much simpler and very different model of *definition, theorem, proof,* as discussed in Moore (1994). Hanna (1995) too has observed that whereas mathematical practice uses proof to justify and verify, mathematics teaching uses it to explain, and Ernest (1998, 1999) similarly emphasizes the social nature of proof and the discursive differences between pedagogic and research mathematics. As Sierpinska (1994) points out, this is an important distinction: there are substantive epistemological differences between proof and explanation. From the point of view of my general theoretical framework, this is by far the more important disjunct, because it involves much more than a skills deficit. Students' lack of exposure to the knowledge construction process results in undergraduates who "exhibit a lack of concern for meaning, a lack of appreciation of proof as a functional tool and an inadequate epistemology" (Alibert & Thomas, 1991, p. 215) and who are outside of the formal proof culture of academic mathematicians (Harel &

Sowder, 1998). Crawford, Gordon, Nicholas, and Prosser (1994) note the related effect that first-year mathematics students see mathematics learning simply as a rote learning task. Thus undergraduate students see proof in an instrumental and performance-related way rather than as an intrinsic component of being a mathematician subscribing to a mathematician's values, assumptions and practices.

Almeida's (2000) study is illuminating in this respect because it directly accesses undergraduate perceptions of proof and proof practices. Responses to statements about proof from first to third year students showed increasing awareness of key features of proof, but students tended to perceive it as an exercise to be undertaken in all circumstances, even when not required, thus suggesting pedagogic enculturation into proving as demonstration and explanation rather than as a means of gaining insight into a problem. Fischbein and Kedem (1982) found similarly that Israeli high school students did not recognize that provision of a proof meant that further example checking was not required. A number of studies suggest, then, that students have fundamentally different ideas about proof from those of their teachers, and that this is a reflection of the particular practices to which they are exposed and the communities that they participate in. Recio and Godino (2001) argue that students' simultaneous membership of a number of social institutions, each with different ways of defending knowledge claims, accounts for their failure to distinguish between appropriate uses of different types of argumentation. Thus they are familiar with empirical inductive proof in scientific situations, and with informal deductive proof in the classroom, but rarely encounter or participate in formal deductive proof. This analysis concords with the observation that pedagogic exposure to proof—as explanation of a finished product—is rather different from its use by mathematicians as part of a creative process replete with blind alleys and false starts and including both formal and informal proof schemes.

That students' epistemological confusions are the result of a disjunct between the practitioner's tacit knowledge and actual practice of proof and the way in which mathematics is taught and portrayed, and a corresponding mismatch of identity and approaches to problem solving, seems highly likely. As Rav (1999, p. 36) points out, "mathematics is a collective art" with proof playing a central role which goes beyond a purely logical-deductive function. This is demonstrated by Weber's (2001) verbal protocol analysis comparison of doctoral and undergraduate students: while undergraduate students did not necessarily lack the syntactic knowledge required for proof, they did lack the strategic knowledge of the doctoral students which enabled them to put this into operation. Thus 57% of failed undergraduate attempts at proof were failures to *apply* syntactic knowledge; when prompted to use this knowledge the students were able to construct proofs. In contrast, the doctoral students employed heuristic knowledge to construct proofs successfully: they "appeared to know the powerful proof techniques in abstract algebra, which theorems are most important, when particular facts and theorems are likely to be useful, and when one should or should not try and prove theorems using symbol manipulation" (p. 101). Thus Weber argues that "an understanding of mathematical proof and a syntactic knowledge of the facts of a domain are not sufficient for one to be a competent theorem prover" (p. 107). Raman (2001, 2003) makes similar distinctions between college students', graduate teaching assistants' and college professors' perceptions of proof, based on the relationship between private and public aspects of proof (i.e. "an argument which engenders understanding" versus "an argument with sufficient rigor for a particular mathematical community" (Raman, 2003, p. 320)). For teachers, these are interlinked, for students they are not. Thus students may have an informal understanding, or they may have a more "top-down" procedural approach which is formally appropriate but carries no understanding; teachers, in contrast, link these two in what Raman calls a "key idea": "an heuristic idea which one can map to a formal proof with appropriate sense of rigor. It links together the public and private domains, and in doing so gives a sense of *understanding* and *conviction*" (p. 323). What is important about this distinction is its implications not only for individual strategy but also on pedagogy:

> For mathematicians, proof is essentially about key ideas; for many students it is not. This is in part because students do not have the key idea (an issue of knowledge), but more interestingly because they do not realize that proof is about key ideas (an issue of epistemology). Further, it seems to be the case that even though mathematicians value key ideas in their own work, they do not tend to emphasize those key ideas in instruction, and more crucially, in assessment. (p. 324)

Thus an analysis of proof presents an interesting case study of the inter-relationship between multiple communities of practice: accessing the academic proof procedure is a question of students' ability to apprehend practices which are at best only implicit and at worst obscured by experiences of teaching which carry another message, coinciding with many students' pre-existing episte-mological frameworks. Their ability to see through this opacity depends on the extent to which they are able, or enabled, to stand back and recognize their position with respect to a particular community of practice, and to develop a participative identity in relation to that set of practices.

Negotiating the Boundaries of Higher Mathematics—Identity and Authority

Cobb and Yackel (1998) describe how the social and socio-mathematical norms of classrooms prescribe teachers' and students' roles and the nature of know-ledge, explanation and justification. In the transition to university it might seem reasonable to look for a shift in identity and perceived authority towards a more autonomous and participatory role for the student. There is little evidence of this, however. On the contrary, there is a strong theme in the students' accounts of disenfranchisement in the learning process, and, for some, adverse comparisons with their prior experience at school. As we have seen in the previous chapter, Diane recounted how her school mathematics experience had been far more participatory and connected than at university, and Carol also commented that learning mathematics did not lend itself to the university set-up. In her view it was the pedagogic structure rather than the content which made it difficult:

> I think there is quite a jump. . . . I think it's going from a classroom to a lecture which is difficult to start with. I think it's hard to lecture maths . . . The tutorials basically were just going through the homework questions. If there was an area of difficulty it would be addressed by the tutor but it was more of a formal setting.

Within the university itself, the students describe themselves as outside of the mathematics community. Their relationships with the lecturers require them to engage only with mathematics as already created rather than with the disciplinary process of creation and validation of knowledge, and their experience of missing explanations and exclusivity places them on the periphery of the community—learning mathematics is out of their control. Sarah, for example, complained about lecturers who challenged too much and, like Seymour and Hewitt's women students, suggested that this was counterproductive:

> And sometimes when they gave you the questions, you always felt like they were trying to make it difficult for you to do . . . [Good lecturers] don't over-complicate it and they teach it in a way that helps you to be able to learn instead of this whole trying to trick you thing . . .

Although they were not necessarily able to articulate the difference, the students effectively described their university teachers as having two types of institutional relationships: their membership of the mathematics research community, and their pedagogic relationship with their students. The first of these was hidden from the students, and based on practices which they were only dimly aware of. The second was their public role as far as the students were concerned, in which they and their students acted in ways which coincided with the students' narrowest epistemic frameworks.

As we have already seen in the last chapter, Sarah's position is an unusual one in the group. Her willingness to experiment demonstrates a different quality of epistemological belief about mathematics learning and mathematics itself. Unlike the other students, she sees herself as a potential contributor to mathematics knowledge, and in a position to engage with the community, albeit in a junior role. She recognizes that it has rules and believes that she has the potential to access and manipulate these rules—despite her beliefs about innate ability reported above. Sarah thus demonstrates an identity of "legitimate peripheral participant" which is lacking in the other students with the possible exception of Larry, who, as we have already seen, looks for usable connections within mathematics. Given the predominant pattern of institutional relationships and their corresponding epistemologies, it is unsurprising that the majority of the students were unclear about the central role of proof. As we shall see, Sarah's outlook led her to take a different and unusual stance on this matter too.

Identity and Authority—the Role of Proof

None of the first-year students demonstrated a very clear understanding of proof and its role in mathematics. Although this may be unremarkable in itself—they simply haven't yet been taught, one could argue—their beliefs about mathematics and their perceived place in it form a powerful context for their comments. Carol presents a common view of proof which reflects many of the findings reported above, as something that lecturers—not students—do, as part of teaching:

> It's something that you get on the OHP when they're doing some new part of pure maths and you're being shown and then you could probably work it out again yourself but I don't think we've ever had to sit there and come up with a reason for the things that we're doing.

Pete says something similar, describing how, like Raman's procedural students, he learns methods but does not understand the proofs:

> I do manage to get the mechanical bits, I don't, I'm not very good at proof, or understanding necessarily but I can learn things and how to do them and apply them.

Joe talks about proof in terms of tutor demonstrations which can be ignored because they are not assessed:

> All the things we look at we're told how to prove it, but then we were told we didn't need to know how to prove it. So as soon as we were told that I never looked at it again, they would show us how to prove it and then they would say "you'll never need this"—so that was that—so I just thought "forget that".

Chris too sees proofs as explanation which can safely be forgotten and believes that they merely serve the function of giving the mathematics department's teaching an appearance of quality:

I'm told "so and so and so and so is this" then I won't go and read and try and understand why. I just remember the result . . . I think they just do it so they don't get criticism of just throwing it at you.

However, while Carol, Joe, Pete, and Chris limit proof to tutor demonstration in Moore's *definition, theorem, proof* sense, Larry and Sue recognize that proof is something which they will have to do themselves as part of mathematics, although they both felt inadequate for the task. Diane feared proof, describing considerable confusion as to its nature and practices:

In the A-Levels the questions I'd always got stuck on would be "show that this equals this". I think, "How am I meant to? What am I meant to use? If I use any of these am I not just using the fact that this is equal to this to prove it?" . . . I see the word "prove" and think "Oh no".

Her negativity about proof as something not only frightening but sterile is also evident in her comparison of mathematics and geography:

I don't know, it's the difference between sort of exercises and essays. In maths you're saying sort of "prove" this and in geography it says "discuss".

Unlike the other students, Charlie, although not explicit on the nature or methods of proof, did associate it with doing "proper" mathematics—by which he meant the use of insight and experience rather than "cookbook" rule following:

I quite enjoy doing that sort of stuff [proof]. . . . A lot of stuff, . . . you either know it or you don't, you just follow a set path and you do it . . . In [proof] you're using maths to do it rather than just your memory.

Charlie's distinction between mathematics and memory indicates his awareness of mathematics as a social practice rather than a collection of algorithms. But although he enjoys proof, he describes mathematicians and their activities as separate from himself and his own interests and capabilities. Steve also placed himself outside of the mathematics community, but in a different way from Charlie: even though he was opting to study a single major in mathematics, he did not see proof as an integral part of doing mathematics. What is striking about his comment here is its illustration of his focus on career and his instrumental view of his university experience:

I concentrate on just doing the methods, I accept whatever it is, I don't question, I don't really focus on proof . . . [*But isn't proof a part of maths?*] I'm not quite sure, I think it would be, it just depends whether it's useful after your degree, it depends what job you do . . . It's interesting to see where it comes from, but I'm not too sure about after, applying proofs.

Debbie, on the other hand, had a far more positive attitude to proof. Her interview illustrates the importance not only of students' epistemologies of mathematics but also their self-positioning as learners. While she did not feel able to engage with proof right now, Debbie anticipated being able to do so:

> It's weird because even though I didn't really understand it, it took me a while to get to understand certain things we have to do, I did sort of feel to myself "I think I'm going to like this". I like the concept of it, I like proof, I think I'm going to be all right with that.

In a way which is predictable from her comments on creativity, Sarah showed the most understanding of the role of proof in mathematics and an eagerness to participate in it. She is willing to try things out as we have already seen; while she recognizes that she does not yet have the skills required she does not interpret this situation as excluding:

> Sometimes, like, I don't mean to but sometimes say if someone is teaching me, or I am doing a problem, I might see, like, a connection between some things and I will think "oh maybe this would work and then maybe I would be able to prove that, and this and the other" which doesn't usually work but, you know, I do think about it . . .

What distinguishes Sarah, and to some extent Debbie, from the other students is the recognition of proof as a part of mathematics rather than as an optional extra. It is as though both seek to grasp the "key idea" as expressed by Raman (2003). Debbie, for example, refuses to "let things go," telling a story of persistence in the face of finding her studies difficult, of how she refused to give up and of how she demanded help from an otherwise aloof tutor:

> . . . I'm still here. There was another mature student that left after seven weeks . . . I sort of thought "well, no, I'm still here, you know, I'm still doing it". So I've got every right to be here . . . I used to knock on the lecturer's door: "please, you know, why with the matrices, what's the point of it?" . . . He was like, "you're gonna have to get sorted out with these type of things, you know," and I flushed up and everything. But I sit it out, you know, because he's upset me before but I just think "no, I'm determined to learn".

Both endeavor to sustain a self-image of potential participant, and in Sarah's case this extends to acting as a participant, albeit in the role of novice or legitimate peripheral participant. Her attitude to proof is part of her wider view of mathematics as a creative process which she can be part of. In this respect her outlook on mathematics is surprising given the practices, beliefs and experiences of the undergraduate community which is so apparent in the interviews in both this chapter and the previous one. Like the women in Seymour and Hewitt's study, she seeks a relationship with her studies which is different from the dominant view; perhaps unlike them, she is managing to maintain this despite the impact of its discourses of speed and ability which she herself voices.

Neither Deficit nor Difference

The case studies reported here are striking in their portrayal of familiar themes when it comes to mathematical epistemologies and as such they provide an insight not only into why undergraduate students struggle with proof but also into their more general self-positioning at university. As the students' comments show, their beliefs about the nature of mathematics as a matter of certainty, rule-following, isolation, abstraction and lack of creativity differ little from those identified by research into school mathematics. Again in correspondence with school research, their beliefs about learning mathematics emphasize speed and fixed ability. Rather than reporting a shift in identity towards perceptions of themselves as negotiators or even *potential* negotiators of mathematical knowledge, these students described themselves as powerless, and university lecturers as more or less aloof authority figures. These beliefs and identities are encapsulated in the specific case of proof: the dominant view was to see it as an irrelevance, or as an unattainable and excluding skill. There is therefore little evidence in the students' accounts of an awareness of the existence of discipline-specific ways of co-constructing knowledge or of their own potential role in accessing these. Indeed, for the most part, the interviews are illustrative of learners who are highly marginalized and, to use Schoenfeld's (1988) terms "passive consumers of others' mathematics." From this point of view, these students are not simply demonstrating a deficit in their skills repertoire in their difficulties with proof. They are working within a different epistemological framework from that of the established mathematics community, a framework which does not include the creativity or ownership associated with proof. Importantly, though, they are not at odds with the public face of mathematics as they experience it within the institution of the university, and in this sense they exhibit neither deficit nor difference—both students and teachers are, it seems, acting in accordance with the same institutional relationships and epistemic frameworks. This is particularly evident in the students' accounts of identity and authority and their corresponding descriptions of proof as mere gloss in the teaching context. Success in these circumstances ultimately relies on students' ability to see beyond this public image and engage in the creative practices which unsurprisingly strike them as so private.

As a number of educators have observed, however, students' experiences of traditional mathematics teaching which emphasizes "received" mathematics are unlikely to engender attitudes and identities which enable them to take control of their own development as mathematicians (Alibert & Thomas, 1991; Boaler & Greeno, 2000). Even where students are encouraged to make their reasoning explicit in an attempt to "make learning experiences more co-operative, more conceptual and more connected" (Dreyfus, 1999, p. 85), lack of clarity on the part of both teachers, students and textbooks as to the relative status of different kinds of mathematical explanations militates against shared epistemologies. As Harel and Sowder (1998, p. 237) point out, "we, their teachers, take for granted what constitutes evidence in their eyes. Rather than gradually refining students' conception of what constitutes evidence and justification in mathematics, we impose on them proof methods and implications rules that in many cases are utterly extraneous to what convinces them." That this is a subtle and participative process involving learning to make connections between "key ideas" and the language of formal mathematical proof is demonstrated by Raman and Zandieh's (submitted) observation of student discussion. In this study of one student's persistence in defending his insight into a problem while neglecting the details of a public justification, Raman and Zandieh argue that conviction that one is right is necessary to motivate the search for a justification, but this needs to connect to argument which demands mathematical evidence. In contrast, teaching which emphasizes mathematics as already created rather than mathematics *in* creation will do little to contribute to this refinement, for the reasons that Schoenfeld (1994, p. 57) outlines:

> When mathematics is taught as received knowledge rather than as something that (a) should fit together meaningfully and (b) should be shared, students neither try to use it for sense-making nor develop a means of communicating with it.

As the analysis of the data shows, an assumption of mathematics as already created, neatly packaged, and accessible only to the quick and able is pervasive and presents a major barrier to the development of a conception of even potential participation in the negotiation, construction or validation of knowledge.

6
Creating Spaces
Identity and Community Beyond the First Year

In Chapters 3, 4 and 5, I have shown that while many learners may be successful in mathematics they nevertheless see themselves as existing only on the margins of the practice, or as lacking stability in it—in this sense, they have what can be called a fragile identity. Although this is by no means the sole province of girls and women, they do appear to express such fragile identities more often or at least more readily. In Chapter 4 in particular, I have observed that an analysis based on communities of practice does not appear to explain the anomaly of women who can be characterized as engaged learners—more engaged than many of their male classmates—and yet have identities of non-participation. I have suggested that these differences are underpinned by discursive positionings which inscribe learners' relationships with mathematics in particular ways, and that student responses to these are gendered—this pattern not only appears in my own data but in that of other researchers, most obviously Seymour and Hewitt's. In this chapter, I explore the range of identities within the under-graduate community of practice in more detail, drawing on the second- and third-year joint interviews and focus groups at Farnden and Middleton universities. As the students talk, joke, and tell stories about being a mathematics student, it is possible to see how they draw on particular, now familiar, discourses in their positioning of self, other students, and tutors.

Identity and Community

In Chapter 3, I drew on Sfard and Prusak (2005) and Gee (2001) in my summing up of the secondary school data and what it indicates about the ways in which students take up particular positions as mathematics learners. Like Gee, Sfard and Prusak suggest that repeated positionings play a major part in the developing narrative of self as a "designated identity":

> On their way into designated identities, tales of one's repeated success are likely to reincarnate into stories of special "aptitude," "gift," or "talent," whereas those of repeated failure evolve into motifs of "slowness," "incapacity," or even "permanent disability". (Sfard & Prusak, 2005, p. 18)

While individuals do have some capacity to choose their designated identities, they cannot act outside of the collective:

> More often than not, however, designated identities are not a matter of deliberate rational choice. A person may be led to endorse certain narratives about herself without realizing that these are "just stories" and that there are alternatives. . . . identities are products of discursive diffusion—of our proclivity to recycle strips of things said by others even if we are unaware of these texts' origins . . . designated identities are products of collective storytelling—of both deliberate molding by others and uncontrollable diffusion of narratives that run in families and communities. (pp. 18–21)

The concept of a designated identity and the emphasis here on collective storytelling seems to be particularly appropriate in the mathematics context. However, it does raise a question as to what options there may be for resistance to, or refusal of, offered positions: Sfard and Prusak suggest that there is (limited) room for choice, while Gee (2001, pp. 116ff) also suggests that it is possible to "bid" for a particular identity position or to resist invitations to take up an ascribed identity (but importantly bidding or resistance are not necessarily successful). Understanding how individuals might exercise agency in this way (or not) is particularly interesting given the mismatch between identities of engagement and identities of participation that I have observed in earlier chapters.

Holland et al.'s (1998) concepts of figured worlds and of the authoring of identity provide a way of exploring the interplay between collective narrative and individual agency in mathematics learning. Together, these ideas not only capture the kind of collective narrative that constructs the mathematics learning community, but they also enable us to see how individuals might draw on particular resources and discourses to re-figure their relationship with mathematics. Within the figured world—"a socially and culturally constructed realm of interpretation in which particular characters and actors are recognized, significance is assigned to certain acts, and particular outcomes are valued over others" (p. 52)—identities are enacted and produced, and individuals take up positions in accordance with "the day-to-day and on-the-ground relations of power, deference, and entitlement, social affiliation and distance" (pp. 127–8). But despite this strong underpinning of positional identities in figured worlds, agency is possible, through reflection on the nature of the figured world itself.

> The everyday aspects of lived identities . . . may be relatively unremarked, unfigured, out of awareness, and so unavailable as a tool for affecting one's own behavior. . . . [But] Ruptures of the taken-for-granted can remove these aspects of positional identities from automatic performance and recognition to commentary and re-cognition. (pp. 140–1)

Although Holland et al. suggest that it is possible for the individual alone to achieve this level of reflection, this is also a collective process:

> This disruption happens on the collective level as well. Some signs of relational identity become objectified, and thus available to reflection and comment ... Alternative figurings may be available for interpreting the everyday, and alternative ways of figuring systems of privilege may be developed in contestations over social arrangements. (pp. 141–2)

Most important for my current purposes is the authoring of identity, described here by Skinner, Valsiner, and Holland (2001):

> The author of a narrative generates novelty by taking a position from which meaning is made – a position that enters a dialogue and takes a particular stance in addressing and answering others and the world ... Thus speaking and authoring a self can be a creative and novel endeavor ... In weaving a narrative, the speaker places herself, her listeners, and those who populate the narrative in certain positions and relations that are figured by larger cultural meanings or worlds. Narrative acts may reinforce or challenge these figured worlds. (para 10)

In what follows, I examine the narratives of doing and being good at mathematics which are generated in the second- and third-year undergraduate focus groups. I will show that they do indeed place themselves and others in particular positions and relations which reinforce the figured world of mathematics. I will suggest too that their stories can also demonstrate a challenge to this world, and a move towards its refiguring, which finds voice in their expression of fragility and instability and at the same time enables them to carry on in what are for many the first time they have encountered difficulties as mathematics learners.

Genius and Geeks—the Figured World of the Undergraduate

The beliefs about mathematics and mathematicians that we have seen in the previous two chapters persist beyond the first year it seems. The second-year students also told stories about students who are able to pick up what a lecturer is saying with little effort, and it is these students who are "good at maths"—here are Tamsin, Roz, and Caitlin (2nd Year, Farnden) on the subject:

Tamsin: Some people are just naturally good at maths ... they just know it straight away.... sometimes someone can put so much effort in and another can put not as much and then not as much can do better.

Roz: There are some people who find at this level all of maths easy, because they can naturally see it and they can intuitively understand it.

Caitlin: For some of them they just pick it up so quickly, so what they can do in the lecture is enough for them to pass the exam. Whereas others of us will have to do all the exercises, and all the past papers and revise really hard to just get the same as they do.

Despite their positioning of some mathematics students as outstanding "others," there is a closing of ranks when they compare themselves with non-mathematics students, united in their "geekiness." Liz and Rachel, Farnden 3rd Years, tell a jokey story of ostracism and cleverness:

Liz: I've actually had someone turn round and not speak to me when I told them what I do.
Rachel: Every time I tell someone that I do a maths degree "you must be really clever" . . .
Liz: They just go—"oo—maths!"
Rachel: If you say you're maths you're geeky . . . it's just the stereotypical thing that comes with maths. . . . I think people are very stereotypical about maths, they just assume you're very geeky, especially if you're at degree level.
Liz: [People say] "I struggled with GCSE."
Rachel: . . . and if you've gone on to do a degree you must be—superhuman!
Liz: Someone did say we should be psychoanalyzed . . .

Although they give a similar account of this difference, Roz, Tamsin, and Caitlin ultimately distance themselves from the geeks:

Caitlin: Non-maths people call us maths geeks . . . the first thing they do [is] pretending to use a calculator . . .
Roz: I think it's just something that comes with doing maths [*laughter*]
Caitlin: But I don't class myself as a maths geek (*Do you class anybody in your year as a maths geek?*) [*much laughter—clearly they do*]
Tamsin: Well they're very dedicated . . . [*more laughter*]
Caitlin: I suppose we all are in a way. I wouldn't class them as geeks but some are more dedicated than others . . .

Another category of mathematics student is the genius. Nick is positioned in this way in his Middleton 2nd year focus group, actively contributing to its construction through his own narrative of his mathematics career. His story is that he had always known that he would study mathematics:

I could have told you in primary school what I was going to do. . . . I just loved doing maths so I never really . . . [*too much noise to hear as they laugh and joke*] even when I was 3 or 4 . . .

The barracking around Nick's admission here is part of the group's positioning of him as "good at maths" and of themselves as less so. Among the second-year students, the belief that mathematics is the province of the super intelligent is evident from their comments when they are asked to talk about those who are good at mathematics and how they deal with being stuck: throughout the focus group we see the construction of Nick as a "genius." Discussing how they have coped with new material, number theory is mentioned; Nick says that this was OK but Megan interjects "you're a genius Nick"—number theory had been problematic for the rest. Later, when I ask what they do when they run into difficulties, Megan responds "ask Nick!," a response which is repeated by the whole group—they laugh and look at Nick. Although this group frequently work together, they claim that Nick has no real need of this. Megan explains: "everyone's good at different bits (Nick is just good). The rest of the group do little bits and we kind of work together." Finally, Liang makes another Nick joke: "Nick needs some intelligent conversation, he's got to talk to himself!"

In comparison with Nick, Jess and Emma position themselves—and tell how they are positioned by others—very differently. Jess talks about aiming high but always being second best:

> I think I'll be over the moon if I get a 2.1 . . . I'll probably get a 2.2 but I always aim for one mark above—I always remember something a GCSE teacher said to me: he said, "if you only need a B to get into the 6th form then aim for an A so if you get a B you won't be disappointed" so you always aim for the highest mark that's going . . . I didn't get A at A-level I only got B, and I didn't get the A* at GCSE I only got A, so I'm not going to get a first I'll only get a second.

Emma also finds it difficult to describe herself as a good student, both to herself and to others. Her identification of herself as able is set within the context of her tutor's assessment of her:

> My tutor seems to have high expectations of me after my results last year but I just hope I get through it and get a decent grade by the end. . . . I used to think [I can't get a first] but last year I got a first so it's kind of a big shock, and that's why I think my tutor has more faith in me than I do. . . . I never had it so to say I'm going to get this and I'm going to get that when I don't actually know, I don't want to say it and then fall flat on my face.

Matt, in the same Middleton focus group as Emma, tells his success story in a strikingly different fashion, his reference to his tutor acting as evidence for his self-positioning rather than as a contradiction to it:

> I don't want to sound big headed but I'm hoping for a first. I think I'm on the way to getting that. . . . my tutor has been trying to get me to do a PhD . . .

The importance for women of relationships with tutors which is described by Seymour and Hewitt seems to be in evidence here also. Emma's performance has earned her more nurturing treatment from her tutor, and this is important to her positioning as a student novice:

> It helps because whenever I go to see him when I'm stuck he doesn't think "well, I'm not going to help her she can figure it out for herself", he's always, like, "I'm going to help her then she can get there".

Jess, however, on hearing this and being asked if anyone else has this sort of experience says, plaintively, "my tutor laughs at me," and—unusually—this is not greeted with laughter, rather, expressions of sympathy. I turn to the part played by tutors in the next section.

Significant Actors: The Tutors

The students' descriptions of their classroom experiences focus on some familiar central themes: narratives ("war stories" as Seymour and Hewitt call them) about the lecturers' behavior in class, both in general and in relation to individual students, and the content and delivery of the various modules. Their tutors are significant actors in these stories, as the comments from Jess, Emma, and Matt above show: they are portrayed as the owners of esoteric information (and they have personalities to match) that the students are only entitled to if they can meet the challenge of second- and third-year pace and increasing independence and so earn their tutors' approval.

The theme of both second- and third-year students' accounts of doing mathematics is "stepping up": dealing with more difficult content and faster delivery. Roz (Farnden, 2nd Year) says that both of these factors are significant, and they have a corresponding effect on the way that students approach the work:

> There is a big step up in the complexity of what we are being taught and the speed at which we are being taught to absorb it and get to grips with it. . . . because it's a lot more complicated, the work takes longer and you can't afford not to do it.

In the lectures themselves, the fast pace of delivery meant that often the students were simply writing notes without having time to think about what they were writing. This appears to contribute to a sense of disempowerment, in contrast with a good lecture, described here by Liz (Farnden, 3rd Year):

> I just understood it, I listened and I really enjoyed it . . . Actually wanting to go and do the homework rather than "I've got to do that homework that I know I'm not going to be able to do".

The idea of increasing independence is central to the students' narratives of what it is to be successful, but this is couched in terms of being able to survive the withdrawal of tutor help rather than a sense of mathematical development on their part. Rachel and Liz (Farnden, 3rd Year) commented on the third-year step-up to greater independence as one which sometimes left them to sink or swim:

Rachel: Some of them seem to think we have to do more ourselves, this year you should try much more yourself. . . . They're less willing to help you out so much.

Liz: . . . this year . . . it's more a case of you do the homework and if you've come out with the answer then you must have got it right . . . they feel we have to stand on our own two feet . . .

The issue of increased independence is a vexed one. All the students said that they knew they were supposed to be more independent as time went on. Some of them interpreted independence as getting on with the work without sticks or carrots: Tim (Middleton) explained—with interjections from Matt—that:

Tim: Last year I felt I was working a lot harder. In the first year we had 2 or 3 pieces a week at least . . .
Matt: And we had more tests as well.
Tim: . . .whereas this year it's a lot more independent and you've got nothing to force you . . .
Matt: No progress check.

Nick (Middleton), on the other hand, expressed a rather more constructive view of independence which matches other aspects of his mathematics identity:

I think you have to put a bit more effort in to understand what you actually go through in the lectures. If you can't understand . . . you have to do some outside work—if it didn't make any sense in the lecture you've got to sit there and try and sort it out. . . . I think you're expected to do more work outside of the course this year than you were last.

However, while Nick appears able to put this strategy into action, the idea of independence was unclear and daunting to many students. At Farnden, Liz and Rachel tell me that a major problem with their third-year project special study is its uncertainty and lack of feedback from their tutor, while at Middleton, Megan says that she is panicking about the project next year. In contrast, Nick and Liang are relishing the greater agency it will bring:

Nick: Well I'm interested in being able to spend time doing something that you're interested in . . . where you've got a lot more time to think about what you're doing rather than writing the first thing that comes into your head.

Megan: I'm dreading it. I'm the other way round I prefer exams.

Liang: In an exam, the knowledge they test is what you can imitate whereas in the project you can express yourself and you can bring out something that is totally new and that is all your knowledge you know so they can actually evaluate the knowledge you know better than just doing an exam. So I think that is much better.

Nick: When you're at school you're basically just trained to pass exams whereas now you've got a chance to actually do something different and do yourself and spend time on rather than just basically looking for marks.

Megan: I'd just rather do exams. I find it a lot easier to have a question do the question . . . I don't want to have to write a lot about it and explain how I've done it. I just like to get the question . . . work though it, do it and it's done.

Generally speaking, then, the students describe themselves as being in a passive relationship with respect to learning mathematics. Tutors are endowed with an authority which appears to be based in discourses of mathematics which portray mathematicians as verging on genius. Indeed, as Lemke has said, science (and I think we can include mathematics in this) is "a special truth that only the superintelligent few can understand" (Lemke, 1990, p. 149). In correspondence with the images of mathematicians that I discussed in Chapter 1, and with the more general "genius and geeks" discourse used by the students themselves, lecturers are portrayed as somewhat maverick individuals who are allowed to break everyday rules of communication. The stories that the students tell about the mathematics lecturers amount almost to caricature, featuring patronizing behavior, dismissal or blatant refusal to acknowledge students' needs. Many stories are told by the students of lecturers wielding considerable interpersonal power. For example, Emma, Jess, Matt, and James (Middleton) jointly tell a story of how knowledge is obscured and withheld, while students' attempts to gain it are dismissed:

Emma: One of our modules, he gave us a handout at the end of term and he goes backwards and forwards through the handout and doesn't go in any logical order. He misses out chunks I really don't know how you can say they're lecture notes.

Jess: He went from page 10 to page 16 and read the title and then said "read it at home" and then skipped a few pages . . .

James: It's as though he's just printed off a section from a textbook and he's just telling us bits and bobs and its not very coherent. In the lecture he says "I wonder what we'll do today" . . .

Matt: He gave a talk, there was no structure to it . . . He just jumped backwards and forwards I didn't know what he was on about.

Jess: In one of his tutorials—he doesn't particularly help you. I said "I don't understand it can you just look at what I've written for this question" and he just looked at it and said "Your answers are strange" and walked away [*they laugh all through this story—they have heard it before*]

Nick, Yu, and Megan (Middleton) tell a similar story about a lecturer whose behavior challenges all but the students who "know the answer":

Nick: If you've got someone who's going to patronize you if you're totally wrong then you'll be reluctant to shout out (I won't mention any names) . . .: a lot less people would be willing to take on the challenge.

Yu: If people know the answer, they've got no fear.

Megan: They pick on you.

Nick: It's just the response you would get if you were to be wrong it would be "how do you not know"—that kind of response.

Megan: "Why don't you know it, it's blatantly obvious, it's simple"—no it's not!

Despite the fact that this is universally acknowledged as poor behavior, the students seem to expect it, and this kind of story will be told with a good deal of laughter. They go on to tell another story of oddity, this time in terms of inability to communicate normally:

James: In analysis it's not exactly the best subject interest-wise but the lecturer seems really nervous all the time . . .

Jess: When he's not looking at the students he looks at the floor or the board . . .

Tim: But he'll never make eye contact . . .

Jess: But he does try to make it interesting and he'll throw in a fact about a historical mathematician or something to make it interesting and keep your interest . . .

Matt: He doesn't seem to be that confident.

By "othering" the lecturers in this way, the students construct an account of their own position in terms of a power balance with knowledge at its center, as illustrated here by Liz:

In some lectures we are willing to ask because of the lecturer. There is one lecturer you can ask anything because half the time he's written it wrong and he's very willing to admit that and you don't mind asking questions with him. . . . There's some of us won't ask certain lecturers and other lecturers we will, some are more approachable than others.

That the balance of power is in favor of the lecturers is clear: Roz describes how a group of students from another class had been integrated into her own, and how their capitulation to the new lecturer's style was grimly inevitable:

The people who came into our group were a lot more chatty and asking questions, because we worked in virtual silence and never asked anything in that particular lecture and then one or two people came in and they were asking questions left right and center and that was quite different because that was the lecturer's style they'd come from. (*Did they keep it up?*) They got quieter.

Safety in Numbers: Refiguring Identity Spaces

In Chapter 5, I described Debbie's challenge to the image of the unapproachable lecturer, in terms of her persistence in demanding help. She tells this story within the framework of her greater confidence as a mature student, and her right to be at the university, drawing on this rather different background in order to re-author and refigure her relationship in this particular community. While hers is a lone voice among the Bradley first years, there is evidence at Farnden and Middleton of a more collective refiguring. These students tell how they work together as a group in ways which are much less individualistic than the discursive positionings which they describe would predict. In their refigured world, they compensate to some extent for the perceived inadequacies of their formal teaching, working towards a collective understanding at their own pace. Part of their strategy involves the use of space, and it is in this respect that they have an edge over the Bradley students: the existence of a mathematics support center in both universities has made a crucial difference. Objectifying the relational positioning of tutors and students, Roz described a subtle shift in power relations when approaching tutors for help on the neutral ground of the mathematics support center as opposed to their own offices:

If you go to their office . . . you know there's a queue of people behind you, they were doing something before you arrived if there wasn't anyone in the queue ahead of you so you feel like you're bothering them, it's their space as well and you're going into *their* office, whereas maths support is neutral ground for everybody . . . it doesn't belong to anybody.

Rachel and Liz added that they behaved differently in the support center, coming with more focus and motivation. More generally, the students described a shift in their approach towards taking more ownership of mathematics, exchanging requests for help and "bouncing ideas off each other," as Jess says:

It's sometimes if there's two of you and you're both struggling but you've both got half of the answer then you work together you can put your half answers together and get the right answer.

Asked how much difference working together makes to the quality of their learning, they explained that it was essential, and that it was the major site of their learning:

Nick and Megan: A lot.

Yu: I think most of the learning is done through helping each other, everyone's got their strength and weakness . . .

Megan: Yes, working through things.

Their group work practices have evolved into alternative physical and virtual spaces, in which they work without the help of lecturers as James (Middleton) and Roz (Farnden) describe:

James: In the [mathematics support] center if there were a few of you on the same course, you'd go to the library and help each other rather than somebody sat there and telling you what to do. If a couple of you work it out together you feel you've accomplished something really.

Roz: Sometimes we use the message board as well to exchange information relevant to the module if there's a piece of information that's relevant for a piece of coursework or that's relevant to everybody, we'll put it up on the message board and have a discussion about it.

While these students describe a highly social set of activities around working on mathematics, the discourses of ability and of mathematics as the pursuit of social isolates still circulate powerfully within their talk. There is a local construction of Nick, Roz, and Rachel as better students, even "genius" as in the case of Nick. Caitlin reports a minority belief that only "thick" students use the center. Other students—Megan and Emma for example—take up positions of being weak in comparison to others and always needing their help, even though they are themselves successful. The tutors are also frequently positioned as stereotypes, unapproachable and dysfunctional in their cleverness, and some students similarly as "geeks." Within these particular relationships, it seems that only those who are marked as high ability are likely to take a more active role or, even more significantly, be *invited* to take one as they are singled out for attention by lecturers as in the case of Emma and Matt.

Making Sense of Fragile Identities

Mathematics learning cultures are a product of deeply embedded pedagogic scripts, gender and ability discourses, and incidents and events which through repetition and narrative combine to make up individual identity positions and learning trajectories. As Boaler and Greeno (2000, p. 171) argue, they lend themselves very well to analysis as figured worlds:

The figured worlds of many mathematics classrooms, particularly those at higher levels, are unusually narrow and ritualistic, leading able students to reject the discipline at a sensitive stage of their identity development. Traditional pedagogies and procedural views of mathematics combine to produce environments in which most students must surrender agency and thought in order to follow pre-determined routines . . . Many students are capable of such practices, but reject them, as they run counter to their developing identification as responsible, thinking agents . . .

The narratives of doing undergraduate mathematics that I have described in this chapter and in Chapters 4 and 5 match this analysis: we see some students accepting loss of agency, more or less happily taking up a position of alignment which is endorsed by the local student community of practice, while others resist it—what happens for these students in terms of the identity positions they then take up has been a major focus of my analysis. As in Boaler and Greeno's study of high school calculus students in contrasting didactic and discussion-based classrooms, I have found that some students find the "received knowledge" aspect of mathematics as they know it unproblematic, while others find it deeply worrisome—they seek connections and understanding but their desire for agency is in tension with the dominant discourses of doing mathematics. I have called these identity positions "fragile identities" because of this tension.

As I observed at the beginning of this chapter, Holland et al. suggest that while individuals develop relational identities in terms of dispositions to act in particular ways, these can be "disrupted" when reflection enables recognition of positional identities which may then be objectified and challenged, and so lead to refiguring:

> The same semiotic mediators, adopted by people to guide their behaviour, that may serve to reproduce structures of privilege and the identities, dominant and subordinate, defined within them, may also work as a potential for liberation from the social environment. . . . When individuals learn about figured worlds and come, in some sense, to identify themselves in those worlds, their participation may include reactions to the treatment they have received as occupants of the positions figured by the worlds. (Holland et al., 1998, p. 143)

As I have suggested above, Roz objectifies the lecturer–student relationship, as does Debbie in her challenge. They act as individuals but disruption may also be collective as it is for the Middleton and Farnden support center students. A function of fragile identities may be, then, that they are part of the process of objectification which enables this refiguring, a site for reflection on how things are and why they are not as they should be. Not all students seem to achieve this level of resistance—as Gee (2001) notes, some learners take up the positions they are offered, and some do not. What determines particular outcomes for particular students may depend on their personal histories-in-progress, as seems to be the case for Debbie and Sarah at Bradley, and Roz at Farnden, whose stories include reflection, challenge, and agency. Perhaps a crucial issue is whether or not they identify themselves as a participant in this world—Jess and Megan do not appear to do so and this may mean that their fragile identities must remain as such—Holland et al. suggest that some of the college women they interviewed had yet to develop their participant skills in their particular figured world of romance, with the result that they were not in a position to develop agency in it. Emma also seems to take a position of fragility despite her membership of what appears to be a powerful collective refiguring force at Middleton. Indeed, while the Farnden and Middleton students' collective action clearly benefits them, it may be the case that their challenge does not go far enough—positional discourses of ability and speed, and use of the gender-marked category of "genius" are still clearly visible in the mathematics that they tell. Fragile identities ultimately depend on the cultural resources available for their resolution into agency:

> The person "makes" herself over into an actor in a cultural world; she may even, over time, reach the point of being able to evoke the world and her sense of herself within it without the immediate presence of others. But the cultural figurings of selves, identities, and the figured worlds that constitute the horizon of their meaning against which they operate, are collective products. One can significantly reorient one's own behavior, and one can even participate in the creation of new figured worlds and their possibilities for new selves, but one can engage in such play only as a part of a collective. . . . The space of authoring, or self-fashioning, remains a social and cultural space, no matter how intimately held it may become. And it remains, more often than not, a contested space, a space of struggle. (Holland et al., 1998, p. 282)

Looking back over the last five chapters, what appears to be the case is that we can trace a developmental pattern which begins in primary school in which particular students are positioned as having high ability and therefore as candidates for engagement in dialogic talk. We see the same pattern in secondary school where again those with high perceived ability are similarly positioned by self and by others as participants in mathematics, leading to a further sorting and self-selection of those students who go on to engage in post-compulsory mathematics. Within this particular group, we see students evolving ways of maintaining ownership despite their experience of university mathematics as difficult and excluding. The stories which mathematics learners have told are therefore ones of increasing exclusion which only the most hardy resist. They are of course the more privileged students as the research tells us (Macrae & Maguire, 2002). If we look as far back as Black's Year 5 pupils, we can see how the interaction of institution and personal background moves towards the privileging of these students. In the next chapter, I will pull together the threads of the three groups of students who have been the subject of this book. I will argue that, while traditional mathematics pedagogy does appear to foster a narrow epistemological stance and a highly performance-oriented level of engagement, particularly in the undergraduate years, it also carries a message, for those who can either hear it or are explicitly told, that in fact there are creative connections to be made. Developing identities of inclusion is therefore a question of making the central meaning-making practices of mathematics visible, available and accessible for all learners. I will argue furthermore, in Chapter 8, that mathematics pedagogy cannot simply rely on collaboration and negotiation between learners, and that the role of teacher as epistemic, rather than social, authority is an important one which needs to be based in particular kinds of personal relationships.

Part II
Developing Inclusion

7
Subject Positions
Explaining Exclusion from Mathematics

The achievement levels of ethnic minority and working-class students in mathematics, together with their under-representation alongside that of female students in post-compulsory mathematics is the focus of extensive research and policy activity. Furthermore, as the previous chapters have shown, even students who choose to study mathematics at undergraduate level are vulnerable in the sense that they frequently express identities of exclusion rather than participation. I have suggested that this may be a product of their particular educational histories and the ways in which they have responded to the ascribed or designated identities carried in repeated discursive positionings. Excluded identities are not merely a question of lack of confidence, persistence or motivation; they are more fundamentally about a disidentification with mathematics in Schoenfeld's (1994) sense of *doing* mathematics, as one who can develop legitimate mathematical ideas within a personal epistemology which enables not only creativity—or "key ideas" in Raman's (2003) sense—but also access to the use of mathematical warrants—that is, agency with regard to mathematical authority. In this chapter, I will look more closely at the issue of how particular mathematical identities develop, and how they serve to exclude certain groups of learners from the possibility of participating in the construction of mathematical knowledge.

Positioning Power: Class, Culture and Gender

What unites much of the research into under-achievement or non-participation in particular social groups is the recognition that mathematics is socially and culturally embedded. From a communities of practice viewpoint, this means that learning school mathematics is a matter of entering into a culture of knowing which is not necessarily self-evident and may be more accessible to some learners than others. In addition, assumptions about the nature of mathematics and the identity positions that these assumptions make available exert further constraints on who does mathematics and how. In this section, I will review research which indicates that traditional mathematics teaching and curricula have the effect of denying many learners access to high-status mathematics knowledge. In particular, it denies them access to meaning-making in mathematics, perpetuating narrow epistemologies, marginalized identities and a corresponding lack of ownership.

Because gender has, although initially unbidden, dominated much of my analysis, I will examine in detail the issue of girls' and women's experience of mathematics learning. In Seymour and Hewitt's work, for example, while both women and men criticized the teaching which they received at college or university level, they responded in different and important ways to the experience, as I described in Chapter 4. One way of explaining women's particular response to mathematics education is in terms of "ways of knowing" (Belenky et al., 1986). Drawing on feminism of difference, this view involves making a major challenge to conceptions of what doing and knowing mathematics entails. I will suggest that this challenge needs to be extended to recognize that, while it may have an individualized and asocial public epistemological face which is reflected in the communities of practice and their associated identity positions that I have explored in earlier chapters, mathematics is—like any body of knowledge—intrinsically social and connected in nature. It is this aspect of mathematical knowledge that needs to be made visible and accessible to *all* learners.

Class and Culture

Issues of class and culture are played out in school mathematics in a number of interrelated and increasingly complex ways, from access to the mathematics register, to the positioning of learners by the kinds of textbooks they are given, to issues of what counts as mathematics, and even to *who* does mathematics. At the level of the mathematics register, for example, Walkerdine (1988, 1997) suggests that while school mathematics makes use of the contrastive pair "more" and "less," in the British working-class domestic talk that she analyzed, quantities were not compared in this way—rather, "more" contrasts with "no more" in the context of mothers' regulation of their children's behavior, notably their consumption of expensive commodities and food. In addition to the novelty of the contrast and its possible implications for understanding, Walkerdine argues that "more" thus signifies something very different from its use in the early mathematics classroom; alongside other similar mismatches this sets up negative relationships with mathematics which are far-reaching (see also Shaw, forthcoming). Taking a different tack, Zevenbergen (2000) argues that students' "linguistic habitus" will affect their ability to deconstruct mathematical texts and the meaning-making that they carry. The prevalence of polysemous (multi-meaning) and homophonic (different spelling but same sound) vocabulary in English and its relationship to the technical lexicon, the specific usage of prepositions and relational terms in mathematics, and its lexical density are factors which she suggests present particular problems for working-class and ethnic minority students, explaining why indigenous students in Queensland performed significantly worse than non-indigenous students on a state-wide mathematics test (Zevenbergen, 2000, p. 207). Similar difficulties arise, she suggests, for working-class English speakers: those who have "rich prior experiences in language . . . are more likely to be able to make sense of and hence construct more appropriate forms of mathematical meaning than their peers who have not had the same rich prior experiences in language. The potential for achievement is greatly enhanced" (Zevenbergen, 2001, p. 44).

Khisty (1995) similarly suggests that lack of explicit access to the mathematics register is the cause of under-performance in Hispanic students' learning in the USA. Languages differ in the way that they express mathematical ideas and in the difficulties they subsequently present for learners—homophonic confusions for example differ between languages—in Spanish "cuarto" means both "room" and "quarter," but there is no corresponding English confusion between these words (p. 287). Importantly though, this is not a simple deficit explanation, and Khisty's research shows that features of classroom discourse and teacher knowledge and beliefs combine together to generate particular student outcomes. Thus a significant issue is that the bilingual teachers in her research tended to use Spanish in the context of general classroom processes (encouragement, class management) but not as part of the construction of mathematical ideas—their general perception was that mathematics is essentially language-free. In conjunction with the fact that despite their bilingualism they also had little access to the Spanish mathematics register, this assumption prevented them from making the register explicit to their students. Thus Khisty argues that "we have operated too long with the myth that mathematics teaching and learning transcends linguistic considerations . . . such a mythology is particularly detrimental to the educational advancement of high-risk students" (p. 295). Moving beyond a focus on the possible restrictions on understanding of such lack of access, Moschkovich (2007a) argues that while bilingual learners may appear to be deficient in terms of their access to the register, they draw on varying resources, including everyday experience, gestures and objects, to communicate in ways which do reflect valued mathematics discourses, and that instruction needs to build on these. Thus a participationist view emphasizes that "bilingual learners may be different than monolinguals but they should not be defined by deficiencies. . . . Bilingual students bring multiple competencies to the classroom" (p. 102).

However, differences also have an impact on the ways in which learners are perceived—as Zevenbergen points out, class- and culture-based language use also mediates access to mathematics via displays of linguistic and cultural capital which mark out an ascribed status of competent learner. Thus, drawing on Bourdieu and Wacquant's (1992) notion of linguistic capital and Bernstein's (1990) concept of restricted and elaborated codes, Zevenbergen argues that the mathematics register "acts as a social filter" (2001, p. 40): "Students who display or assimilate those socially legitimated linguistic practices within their own repertoire of behaviors are positioned more favorably. . . . To be constructed as an effective learner of mathematics, students must be able to display a competence with these forms of texts" (2000, pp. 203–4). Such displays of capital are crucial in terms of future access. Teacher perception of ability is predictive of where a pupil is placed in the setting hierarchy in British schools: for example, Gillborn and Mirza (2000) record that black students are very likely to be allocated to lower-tier GCSE sets, as are working-class students (Gillborn & Youdell, 2000). An extensive study of higher attainers based on data from 2001, 2003, 2004, and 2006 in England suggests too that pupils eligible for free school meals (a standard official indicator of socio-economic status in the UK) or living in deprived areas are less likely to be high attainers at Key Stage 4, compared with other pupils with similar prior attainment (Government Statistical Service, 2007). Significantly, the odds of being entered for higher tier (Level 6–8) papers at Key Stage 3 are lower for these pupils, even allowing for prior attainment. These classroom mathematics trajectories are produced in interaction with pupil self-positioning which itself draws on capital resources: Noyes (2007) demonstrates this in his comparison of two primary school children, one middle class and one working class, and their family-based narratives. In the same class at school for six years, and with the same teachers, these two children draw on very different resources in their self-positioning with respect to mathematics. Working-class Stacey describes herself as hating mathematics, as "dumb" and "in need of desperate help" (she also has "remedial" lessons). She believes that the teacher dislikes her, and would only change this opinion if she were (impossibly) to become like Matt, the best mathematics performer in the class. Her family narrative is one of not doing well at school in general, and in mathematics in particular. Middle-class Edward, in contrast, receives plenty of practical support at home, where mathematics is highly valued and seen as a source of interest and future success. Like Stacey, he is aware of student rankings within their classroom and the value of speed and competition, but unlike her he draws on cultural resources to position himself as potentially "as good as Matt."

Even within heterogeneous teaching contexts, cultural and linguistic capital display may lead to particular teacher perceptions of pupil capability and correspondingly different treatment, as Black's research shows. Reminiscent of the issues raised by Horn (2007) in Chapter 2, Ladson-Billings (1995) demonstrates a similar effect in her observation of a teacher's perception of African American students as not being equipped to receive anything other than purely didactic instruction: "She cannot apprentice these students because she does not believe that they have anything on which to build" (p. 138). Stinson (2006) too identifies the deficiency discourses invoked in explanations of poorer mathematics outcomes for African American male students which position learners as unable to benefit from reform-based teaching. Citing Strutchens (2000), he argues that "the directive, controlling, and debilitating pedagogy (i.e., pedagogy of poverty) typically faced by African American students sharply contrasts with the types of teaching advocated by the National Council of Teachers of Mathematics" (p. 486). Stinson's own study focuses on discursive positionings and agency in four young, mathematically successful, African American men, and their "ability to accommodate, reconfigure, or resist the available socio-cultural discourses that surround African American males" (p. 478). He remarks (p. 479) that "present throughout each participant's responses ... was recognition of himself as a 'discursive formation' ... who could, and did, actively negotiate sociocultural discourses as a means to subversively repeat ... his constituted 'raced' self." Similarly, Martin (2007) describes how, for African American students, "failure is constructed as normative" (p. 149); the construction of counter-narratives by adult students looking back on their raced mathematics education is central to their positive re-figuring of themselves as mathematics students. However, without backgrounds which demonstrate that succeeding in mathematics is worthwhile, it is difficult to fund such counter-narratives: "life experience as an African American, often characterized by struggle and social devaluation, makes it difficult to maintain a positive identity in the pursuit of mathematics knowledge" (p. 157).

Even with such a positive identity at high school level, undergraduate students frequently find further study a struggle: Seymour and Hewitt's (1997) analysis of SME undergraduates found that non-white students were more likely not only to switch majors, but also to drop out altogether. A number of issues were associated with this pattern, including objective economic difficulties, family responsibilities, the stresses of acting as a role model or taking on an obligation to serve the community, but many centered on their experience of the transition to university. While, as we have already seen in Chapter 4, the "straight lecture style" did not work particularly well for students of *any* subgroup, these students, like many women students, frequently drew particularly negative contrasts between university teaching and their high school teaching. Students who had attended high schools where their ethnic group had been in a majority and/or which had emphasized individual attention, reward for effort rather than performance, and fostering relationships to motivate or retain students, were most at risk. Seymour and Hewitt report that "they responded to their first experiences of 'objective' grading in S.M.E. classes by defining S.M.E. faculty as unfeeling or discriminatory" (p. 335). Like the students in Martin's study, black and Hispanic students reported negative stereotyping, especially with respect to IQ, and some clearly internalized these stereotypes: even those who were about to graduate reported feelings of inadequacy. Those who had persisted had in every case found a student support group which enabled them to resist internalizing stereotypes—but this was not necessarily easy to do for two reasons: those students who were used to individual attention at high school were unused to forming peer groups for support, while peer group norms of self-reliance and independence, together with a lack of peer support for success (as opposed to failure) made it difficult for students to sustain a positive SME identity.

As I have already noted in Chapter 2, membership of different sets in England and Wales entails exposure to significantly different curricula: the GCSE tier system restricts access to students in lower sets in terms of breadth, approach and a cap on the possible GCSE examination marks (Boaler & Wiliam, 2001). This fact alone contributes to a positioning which, although resisted by some pupils—"Obviously we're not the cleverest, we're group 5, but still—it's still maths, we're still in year 9, we've still got to learn" (Boaler & Wiliam, 2001, p. 91)—traps them in terms of an ascribed identity as "not good at maths." A specific effect of this practice is the positioning of learners through textbooks, as Dowling (1998, 2001) shows. An example is the contrast between textbooks written for lower-tier and higher-tier students: while both draw on a mathematized real world (in the example below, estimating car speed from car accident skid marks) as a vehicle for teaching mathematical principles, they differ in terms of how much these purely linguistic esoteric principles are foregrounded as the true target: "only in the esoteric domain are mathematical principles fully discursively available for generalization to other mathematical and non-mathematical contexts. The context-dependency of the public domain [i.e. the mathematized real-world problem] more or less severely delimits its range of application" (p. 189). Thus Dowling argues that standard textbooks written for the lower tier can compound lack of access: they have a low percentage of esoteric text (as we might expect, given the basic assumption that lower attaining students will respond best to deeply embedded problems) but at the same time retain the high status of formal mathematical principles which are under the control of the teacher, not the students. While this is a question of positioning via access to discursive practices, there is a further effect of these textual differences in terms of how the learner is positioned as an actor in the mathematical world. The lack of esoteric domain text and the consequent obscurity of the real underlying mathematical principles at stake serve to compound exclusion in the way in which mathematical activity is framed. Thus in an example concerning the police formula for estimating vehicle speeds by measuring skid mark length, the higher-tier treatment has explicit mathematizing and formulae use—the activity is *clearly* just a vehicle for the mathematics, while the lower-tier treatment emphasizes the *actual* police task, supporting the mathematical principles involved. This is what Dowling has to say about this contrast:

> The [low ability] text presents the task as *for* the police activity: its purpose is to enable the student to be a potential practitioner in the public domain setting. The [high ability] task is clearly *for* mathematics operating in the esoteric domain . . . the student is not being apprenticed to a police officer. Rather the student is, metaphorically, standing alongside the textbook author, casting a mathematical gaze on the non-mathematical activity. This student is apprenticed to the mathematician. (p. 194)

Just as Dowling's contrasting texts position the learner as apprentice mathematician or rule-following technician, so the use of particular contexts as vehicles for mathematics teaching interact with class and culture to position the learner as competent or not. Cooper (2001) suggests that the "realistic" setting of many mathematics problems is more likely to be obscure to working-class children because they lack cultural capital: "the sense of the 'obvious' or the 'appropriate' has to be learned, either in the home or the school ... opportunities for learning what is appropriate in school mathematics may not be equally distributed across social class cultures ..." (p. 248). He cites data to suggest that working-class pupils are more likely to import "inappropriate" everyday knowledge into their solutions to "realistic" problems, so falling foul of the (unwritten) rules that problems which refer to real-life events should nevertheless be answered with proper regard to their mathematical analogues, at the expense of considerations of what is and is not likely in real life. Ironically, and in line with Dowling's observations, Cooper also notes that SATs tests in England and Wales at age 14 are more likely to carry questions embedded in realistic contexts at the lower tier levels—where working-class children are more likely to be found. The class effect is best illustrated with an example from Cooper (2001, p. 253) in which Key Stage 2 (ages 8–11) pupils are asked to "organise a mixed doubles tennis competition" by pairing boys and girls with names drawn from two bags—one containing three boys' names and the other containing three girls' names. They are asked to "find all the possible ways that boys and girls can be paired." Cooper reports a significant class difference in performance on this question: working-class pupils were more likely to treat the problem "realistically" (thus finding only three pairs) than were middle-class children, who were more likely to home in on the mathematical interpretation—i.e. the production of nine pairs. It might be argued that a lack of content familiarity could be responsible for this pattern—working-class unfamiliarity with mixed doubles tennis could present a problem. However, the pupils solve the problem on a realistic level without hesitation, failing to see beyond this to the esoteric domain which lies behind. In this sense, content is not the issue: no amount of cultural relevance—turning the question into the Football Association Cup draw, to take Cooper's example—will make the situation for working-class pupils any better.

Similarly, Lubienski (2007) found that seventh-grade lower socio-economic-status students in her NCTM *Standards*-based classroom regretted the loss of pages of decontextualized problems from their mathematics diet, finding their new contextualized problems a struggle. They too "tended to consider a complex variety of real-world variables in solving the problems. Yet in the process, the students sometimes approached the problems in ways that allow them to miss the generalized mathematical point intended by me and by the text. The higher-SES students were more likely to approach the problems with an eye towards the intended, overarching mathematical ideas" (p. 18). So, for example, faced with a fractions lesson based on a pizza-sharing problem, lower SES students focused on shares for late-comers to the restaurant, and options on second helpings. The social class contrast is sharply drawn in pupils' overall responses to their lessons: "whereas more higher-SES students complained about seeing the same mathematical ideas repeatedly with different storylines attached, more lower-SES students complained that they did not know what mathematics they were supposed to be learning" (p. 18). Ladson-Billings (1995) gives a similar example, this time of inner-city African American students' responses to the problem of which is cheaper—a daily return bus fare of $1.50 or a monthly pass for $65. White middle-class suburban students saw the answer as unproblematic—it was clearly cheaper to pay a daily fare on the assumption that going to work for a month entails 20 two-way journeys; African American students on the other hand asked for more information based on a complex real-world knowledge—the traveler may have multiple jobs, they probably do not own a car so may need to use the bus for other reasons, and so on.

Taking one more step back, we need to problematize the relationship between "informal" and "formal" mathematics. The question of "whose mathematics?" and the implications of the answer for the positioning and self-positioning of learners forces a return to the issue of context. While, as we have seen, curriculum design often makes the (questionable) assumption that particular "everyday" contexts will smooth the way into formal mathematics, we need to separate out the use of "context" as a vehicle for teaching school mathematics, versus its role in different kinds of mathematical activities. Carraher and Schliemann (2002) emphasize this distinction when they conclude that:

> The outstanding virtue of out-of-school situations lies not in their realism but rather in their meaningfulness. Mathematics can and must engage students in situations that are both realistic and unrealistic from the student's point of view. But meaningfulness would seem to merit consistent prominence in the pedagogical repertoire. One of the ways in which everyday mathematics research has helped in this regard has been to document the variety of ways in which people represent and solve problems through self-invented means or through methods commonly used in special settings. (p. 151)

So, for example, research by Lave, Murtaugh, and de la Rocha (1984) and Nunes, Schliemann, and Carraher (1993) found that mathematics users in a Californian supermarket and on the streets of Recife, Brazil respectively performed significantly better in these situations than on pencil-and-paper tests, and were more likely to try to make sure that the solution that they arrived at made sense in out-of-school contexts. As Carraher and Schliemann point out, meaningfulness is key; thus research such as Nasir's (2007) work on "basketball mathematics" finds that this out-of-school context supports and validates African American students' identification with the practice of mathematics as they "became more engaged with the sport, ... became more skilled with statistical calculations and, in turn, became even more engaged with the basketball community and established stronger identities as 'ballers'" (p. 143). However, school mathematics is highly valorized compared with the use of mathematics elsewhere. Unsurprisingly, then, De Abreu and Cline (2003, 2007) found that children from marginalized social groups are aware that their own "out-of-school" mathematics is not legitimized by the school. Working with children from a Brazilian farming community and an inner city multi-ethnic area in England, they report that neither group recognized people from their own community (farmers and taxi drivers) as "users of mathematics." As such, these children accepted an ascribed identity which perceived their community as "illiterate/innumerate"—just as Khisty, de Abreu, and Cline identify a damaging view among teachers that mathematics is culture-free and universal, and hence not subject to potential inequities, being only predicated on "ability."

Kassem (2001) also notes the implications of the assumption of value- and culture-freedom in mathematics and its contribution to the invisibility of ethnic minority communities and their disidentification with mathematics. He argues that, even when issues of minority ethnic attainment in mathematics have been addressed, multicultural approaches have involved a bolt-on approach to cultural diversity, using illustrative resources drawn from other cultures such as Rangoli patterns to teach symmetry or Islamic tessellations, but in isolation from their history and meaning in the cultures in which they evolved. The approach of "Saris, Samosas and Steel bands" (p. 68), paralleled by Ladson-Billings' (1995) observation that "what passes for multicultural education in many schools and classrooms is the equivalent of food, fiestas, and festivals" (p. 127), does not address minority ethnic exclusion because it does not consider the role of mathematics in the world and our relation to it: it does not enable genuine participation in the practice of mathematics. Ladson-Billings similarly argues that students need to see themselves as potential contributors to mathematics, citing examples of a Los Angeles teacher's emphasis on the mathematical discoveries of her Latino students' Mayan ancestors, and a teacher of African American students' lessons on the African origins of algebra. With respect to social class, Frankenstein (1995) argues for a mathematics curriculum which addresses how real-life class inequities persist in mathematical terms and how mathematics is frequently used to underpin them; similarly, for Gutiérrez (2007), a "critical mathematics takes students' cultural identities and builds mathematics around them in ways that address social and political issues in society, especially highlighting the perspectives of marginalized groups" (p. 40).

What is crucial in many of these researchers' findings is a focus not just on curriculum change but also on a pedagogical change to a more negotiated mathematics which enables identification with the subject. Gutiérrez summarizes the issues for the development of identities of inclusion in her observation that in the underlying deficit theory of much research "the assumption is that certain people gain from having mathematics in their lives, as opposed to the idea that the field of mathematics will gain from having these people participate" (p. 37). I return to these issues in the final chapter.

Gender and Mathematical Knowledge

In England and Wales, although girls are now matching the performance of boys, they are slightly under-represented in the top 5% of scores at 16 years old (Government Statistical Service, 2007), and post-16 are significantly less likely to take up mathematics, although those girls who do take mathematics at A-level (21,215 in 2007 compared with 32,245 boys) are more successful—45.2% scored at grade A compared to 42.8% of boys in 2007 (Government Statistical Service, 2008). In the same year more than twice as many boys than girls took the Further Mathematics A-level, with slightly more boys than girls scoring an A grade (57% versus 56.6%). The most recent statistics for combined undergraduate and graduate level study of mathematics show 9,915 women versus 16,845 men in 2005/6 (Higher Education Statistics Agency, 2008). In the USA, boys scored higher than girls at ages 13 and 17 in mathematics in 2004 (IES, 2006) and in fourth and eighth grade in 2007 (Lee, Grigg, & Dion, 2007). Take-up of post-secondary mathematics is increasing but still lower for women (Clewell & Campbell, 2002), while men continue to outnumber women with respect to enrollment in high school advanced mathematics courses, in a pattern which appears to have changed little between 1982 and 2004 (Dalton, Ingels, Downing, & Bozick, 2007). These statistics are important because they show that gender issues in mathematics tend to revolve around under-participation, not under-achievement. I have already explored some of the literature on girls' and women's responses to school mathematics in Chapters 2 and 4 and I have also suggested that there are gender differences in my own data; here I will focus on research which theorizes these relationships and addresses the place of gender differences within the broader equity and reform agenda.

A view which has gained some popularity in recent years is that girls and boys, and women and men, relate to mathematics knowledge in fundamentally different ways. Thus Becker (1995), drawing on Belenky et al.'s (1986) seminal *Women's Ways of Knowing*, suggests an explanation of patterns of women's participation in terms of gender differences in preferred ways of knowing and working which result in their exclusion from mathematics, even when they are good at it. She argues that traditional mathematics teaching favors the (stereo)typically male approach of "separate" as opposed to female "connected" knowledge (Gilligan, 1993), thus constructing mathematics knowledge in a way which ultimately alienates and disempowers girls: the "male model for knowing that in its very formation, excluded women, denied their truths, and made them doubt their intellectual competence" (p. 164). Beginning with an epistemological typology which shows similarities with Perry's much earlier general formulation, she suggests that students' responses to mathematics range from silent uncritical acceptance, through "received knowing" which relies on appeal to authority, through a more subjective knowing which is based on a feel for how things are, to knowledge which is procedural and based in method and, finally, constructed knowledge, which integrates intuition and the wider knowledge-base of the community—"it is in this stage that the learner can integrate her rational and emotive thoughts" (p. 168).

In many respects, this typology maps on to other analyses that I have described—we see perhaps Wenger's aligned learner in the procedural type, the engaged learner in the constructed type. Raman's "key idea" is potentially observable in the subjective knowing type. Becker's important contribution to the debate, however, is the distinction she draws between male and female responses within the subjective and procedural knowing types. Thus in the state of subjective knowing, men and women differ in the warrants they present for their intuition: for men, what they know is "obvious" (perhaps the certainty expressed by Raman and Zandieh's persistent student fits this characterization), while women acknowledge that their view is "just their personal one." In the state of procedural knowing the important contrast is between "separate" and "connected" knowing. Becker suggests that men typically exhibit "separate knowing," which emphasizes "the use of impersonal procedures to establish truths . . . The goal of separate knowing is to be absolutely certain of what is true" (p. 166). Although the separate knower by definition will proceed to do this by following the rules of the relevant discourse—in our case those of mathematics—they will according to Becker engage in this in a way which "often takes an adversarial form which is particularly difficult for girls and women, and separate knowers often employ rhetoric as if playing a game" (p. 166). In contrast, the typically female form of procedural knowing is "connected knowing," which is based in experience and "explores what actions and thoughts lead to the perception that something is known . . . authority derives from shared experiences, not from power or status" (p. 166).

To what extent might this typology explain the data I have presented in earlier chapters? Clearly, some of the undergraduate women I interviewed express a dissatisfaction with not being able to fully understand the reasoning behind the mathematics that they are learning, or to really participate in it, and this might map on to the idea of connected knowing. However, this is also true for some of the men—as Seymour and Hewitt also found, they are critical of the mathematics which is presented to them. What does seem to distinguish men and women is their *response* to this situation—the identity that they take up in the process—but this is in interaction with the ways in which they are discursively positioned within this particular world, not a result of their individual characteristics. As Leder (1995) argues, "gender differences in mathematics learning are often a reflection of prevailing circumstances rather than an indication of absolute and unchangeable differences between males and females" (p. 209). So, although Becker's analysis provides a starting point for considering how learners respond to mathematics teaching, its emphasis on "ways of knowing" as stable and gender-specific characteristics underplays the ways in which individuals "negotiate a sense of self," as Boaler and Greeno put it—in their study of high school calculus students, "learners seemed able to move in and out of different "forms of knowing" in different circumstances (p. 190). Taking this idea one step further, as I have suggested in Chapter 6 in particular, we can see learners as being in a process of self-authoring within a practice which they may or may not objectify and challenge—it is this self-authoring that is crucial in their developing relationships with mathematics. Bartholomew and Rodd (2003, pp. 11–12) make a similar point when they argue that:

> . . . we do not regard these women's behaviour as reflecting a truth about them, but rather as arising through a complex interaction of the indivi- dual and the wider context, whereby structures are constituted by the actions of agents, but action itself is organised within the parameters of existing structures.

In investigating this interaction, one of the major points that I want to make is that there are particular discourses within and about school mathematics which serve to hide and exclude by setting up constraints on available identities—for example, it is only possible to be deemed "good at maths" if you are very quick and appear to find it easy, and this precludes a whole range of activities—asking questions, say. And yet, as we have seen in the primary and secondary school data, some students do get to ask questions and to discuss and to make connections. They have privileged access to experiences which support what Becker calls "constructed knowing," in which "the artificial dichotomy between separate and connected knowing . . . becomes apparent and the two ways of knowing can be merged . . ." (p. 168). So when we *engage* with mathematics in this way, we combine personal intuition or insight and the rules of the practice—they are complementary. Together, these two actions enable participation through the use of publicly accountable mathematical pro- cesses which fall under the concept of separate knowing to *validate* know- ledge claims—logic, rigor, abstraction, axiomatics, deduction, structure, and formality—and processes which fall under that of connected knowing to *develop* knowledge—intuition, creativity, hypothesizing, conjecture. To really engage with mathematics, we need both; without the first, we are alone, and without the second we cannot progress. Becker is right about this; so when she introduces the idea of separate and connected knowing as essentially male and female, she is making a distinction which potentially compounds the problem. As Mendick (2005c) argues:

As well as fixing gender, separation/connection fixes mathematics. Separate and connected ways of doing mathematics slide into separate and connected versions of mathematics itself with the former preserved as a space free from the taint of connection. So, the introduction of the idea of connection, by positing the existence of separate mathematics, legitimates what it seeks to challenge. (p. 164)

Thus instead of saying that the dominant projection of mathematics as "separate" (i.e. as privileging the abstract and formal) prevents girls' and women's participation because it devalues their ways of knowing, we could argue that the invisibility of "connected" mathematical processes (i.e. intuition and hypothesizing) and/or their disconnection from the ways in which we can validate our intuitions, disempowers all learners. As we have seen in relation to mathematics and class/culture, what actually lies beneath the public "separated" face of school mathematics is an esoteric domain of pattern finding, hypothesizing and creativity. Although it is largely implicit, it may be revealed to some students who may also be supported in its use, and who take up available identities which enable its use. It is in this respect that girls and women—and indeed working-class and ethnic minority students—may be excluded. As Secada (1995) argues, an over-emphasis on groups and the differences between them tends to obscure issues which are relevant to all. By way of illustration, Leder (1995) describes Rogers' (1990) investigation of a particularly successful mathematics lecturer in a North American university who was male, described as intimidating, and yet attracted a large number of students to his course including women. Rogers reports that the important difference between this teacher's class and others was that he did not lecture, instead encouraging students to work together on problems in small groups. Leder concludes that gender is not the issue here, as opposed to "creating a classroom environment open and supportive for all students—an environment in which the teaching style mirrored the nature of mathematical inquiry" (p. 217).

Like Leder, Boaler (2007) cautions against equity research which confirms stereotypes rather than challenging them, making the point that we if are to improve access to mathematics then we must not locate differences in girls and women but see them as differences which emerge in response to environments. Describing informal research with women university students, she reports a change in self-perception which is very familiar:

In school they had been encouraged and enabled to understand the mathematics they met. But when they arrived at university they found that they were expected to copy down endless formulas and procedures from chalkboards . . . connected knowing may be less accurately represented as a characteristic of women, as it has been in Gilligan's work, than a response to certain learning situations. (p. 31)

She also argues that we are in danger of compounding girls' and women's exclusion from mathematics if we label them as essentially unlikely to perform well in the subject, thus contributing to a discourse of girls' inability in mathematics which, if we are not careful, those same girls will take up in their own self-positioning.

So, while we may agree with Becker's idea of an ideal pedagogy as one which encourages the development of constructed knowing, we might not agree with her reasoning as to why gender differences in participation in mathematics have come about. The classroom context of mathematics teaching is more complex than she suggests, and so is the development of any individual's relationship to a social practice. As I have shown in previous chapters, the interplay of structure and agency over time works to create situations where individuals are positioned, and position themselves, in particular ways with respect to mathematics. In the next section of this chapter, I will re-visit the question of how this happens.

Looking Back Through the Data—How Do Mathematical Moments Become Mathematical Lives?

Becker's analysis suggests that traditional teaching of mathematics presents only one face to the learner—that of a highly valued "separate" knower—and so denies the importance of "connected" ways of knowing. In this section, I want to suggest that, while traditional mathematics teaching does indeed present such a public face, this is not to say that more creative activity—that is, "constructed" knowing—never takes place in the standard classroom: it does. The crucial difference is that traditional teaching does not make this explicit, maintaining instead a surface air of "separate knowing." Success relies on learning which is never explicitly acknowledged, thus creating marginalized identities in those who do not recognize it or who are not helped to see it or who may be prevented from using it. In this section, I want to look again at the mechanisms of disempowerment which result in the exclusion from mathematics of a wide range of learners, pulling together the themes of earlier chapters. Looking back through the data, I will argue that the development of relationships with mathematics entails a complex connection between discourse and identity.

I begin with the undergraduate students, because they are a group who have persisted with mathematics and succeeded in it in the past. As we have seen, many carry on at undergraduate level simply because they can, although undergraduate work begins to present a challenge sooner or later for most if not all of them. Within this group, we see some students who position themselves, and are positioned (by fellow students or by tutors), as "good at maths." Some cite rather performance-oriented relationships with mathematics which could be described as "separate knowing," but others indicate that they are rather more engaged or "constructed" knowers, or at least see themselves as potentially so. Thus at Middleton, as we saw in Chapter 6, Liang and Nick are looking forward to doing project work, because it allows creativity, an expression of self and also a demonstration of deep rather than surface understanding. In sharp contrast, Bradley student Richard, who we saw in Chapters 4 and 5, is very much focused on marks and the "right answers" of mathematics. He has no difficulty in saying that he is good at mathematics, although this relies very much on his marks and being quick to grasp what is going on in comparison with others:

> I do like pure maths, I do enjoy it but stats is less effort . . . when I see a stats problem I can see it quicker than most because I've noticed that in the workshops that I understand quicker than other people on the course—that's one reason why I'm changing [to a maths major] because . . . I have realised I am good at it without sounding big headed. . . . I always knew I was good at maths but I wasn't prepared enough to go to that higher level I didn't think. It was a bog standard comprehensive I was at. The next best A level was a C and I got an A so I didn't know how good other people were at maths and whether I was better or not but looking at the results I think I came 3rd in [one of the first-year mathematics units] so that's when I realised I was quite good.

Richard is only prepared to risk taking mathematics when he is assured of high marks, and will only continue to do so while he is one of the best—as he says, "if I was struggling, I'd drop it." In comparison with the women of Chapter 4—notably Sarah, Debbie, Sue, and Diane—Richard is extremely confident and comfortable with his "separate knower" approach to mathematics. However, it is not clear that any of these women earn the label of a "connected knower," and as I have suggested in Chapter 4, they are more properly understood as students who have seen through the "separate" public face of mathematics to the community of practice that lies behind it. So, for instance, Diane presents a complex picture of her relationship to mathematics, expressing a certain amount of dissatisfaction with her university-level study which manifests itself in an account which, if we apply Becker's typology, expresses constructed knowing. The important point to note is that Diane recognizes that developing mathematically is about moving beyond the exercise of bare procedures towards connecting them in a more creative process:

> . . . in A-level you are told, most of the time you're told "use this, use the product to show this and use this". . . at this stage it's so much more about understanding maths as a whole as in bringing everything together and using everything to work out something.

It is also significant that Diane doesn't see herself as being able to do this. Despite this insight into the nature of mathematics (and the fact that she is studying at degree level), she has an identity of being low in ability. Given her happy memories of school, of sitting round tables working together on understanding, it appears that, like many of the students in the SEUM study and Seymour and Hewitt's switchers, her experience of university mathematics seems to have eroded her perception of herself as good at mathematics, making her vulnerable to the emphasis on individuality and speed and the position of slow and unable that this creates—she cannot "integrate and differentiate all over the place," and so "that's why I think I'm not good at maths." Sarah, who we have seen tell an even more "constructed" story of her creative relationship with mathematics, also has a rather negative view of herself, while Emma, at Middleton, despite being "taken up" by her tutor, lacks confidence in the idea that she could continue with her first-class performance.

There does appear to be a gender difference here in terms of identities of inclusion and exclusion, but it does not fit into Becker's framework. As I have argued in Chapters 4 and 5, identities of belonging do not line up neatly with personal epistemologies and identities of engagement or alignment. We can see here students who could be called "separate knowers"—perhaps Richard is aptly described in these terms since "in accepting authorities'" standards, separate knowers make themselves vulnerable to their criticism. The authorities have a right to find fault with the reasoning of separate knowers; and since there is nothing personal in their criticism, the separate knowers must accept it with equanimity" (Belenky et al., 1986, p. 107). We can also see "constructed knowers" or would-be constructed knowers perhaps: in explaining her liking for mathematics, Sarah tells how she enjoys being creative and exercising intuition, but she also describes how she attempts to use formal proof in order to assess her creation. Nick and Liang are looking forward to some self-expression and sustained exploration of a topic in their projects. What seems important is the fact that these successful students seem to feel differently about their mathematical prospects along a gender divide. Why do the women describe themselves as not belonging so much more often than the men?

Turning next to the secondary school data, we saw in Chapter 3 that a major feature of the secondary school students' accounts were the different epistemologies of mathematics expressed by students in different sets. Many top-set students described how they favored doing "investigations" and how this often engendered different ways of doing mathematics compared with their normal routine of learning rules and procedures. In so doing, many boys might be labeled as constructed knowers, seeking understanding rather than just "right answers," appreciating the opportunity to work with others, and seeking to hypothesize and experiment. In trying to explain the benefits of working in a group, Year 7 Jake described doing mathematics as creative, emphasizing its lack of "straight answers." Here he talks about linking intuitions to achieve a better warrant for a solution:

In maths you can have opinions . . . Because if it was like a puzzle or something you could have an opinion because it could be done two different ways or maybe they might spot something you haven't spotted and you say one thing and they say the other. And so then if you like work together you can put two together and get a stronger answer.

Year 9 Michael picks out another feature of doing group work in his description of the change in relationship with the teacher from authority figure to resource, responsive to pupils' requests for assistance in her role as expert practitioner. Luke and Daniel, in the same top set, talked about the importance of making links between different areas of mathematics. Christopher, in Year 10, provided an example of this as he described how in his investigation he had tried to deal with examples in three dimensions which he found difficult, but he had enjoyed the challenge and the opportunity for contextualization of the mathematics:

I find it challenging but I do enjoy it if I understand what I'm doing. . . . I think investigations are good because you're kind of learning stuff about the world kind of around you, it can be like put into different situations.

As we have seen in Chapter 3, Daniel makes the point that students in higher sets are encouraged to make such links in a constructive way:

. . . we do more of, little bits of more things whereas the people who are lower down do more things with little bits so they don't see as much . . . we sort of see it, we sort of see *all* the maths problems and how they connect to each other and we understand it more . . . but the other people, they don't understand the more complex things and how they fit into each other.

Luke sums this up very neatly in what could be called a contrast between separate and constructed knowing as "the difference between being able to do maths and being able to do the maths investigations."

This is a strong and significant contrast, and as I showed in Chapter 3, it appears to be borne out by the interviews with the students in the lower sets. So Year 9 Trevor describes a "separate" mathematics in terms of performance and memory, and, reminiscent of the textbook contrasts drawn by Dowling, describes how he will be able to use it as a truck driver. Trevor also describes a different rationale for pair work, seeing it more as an error check, and he also hints at an institutional concern about cheating:

I like working in pairs because to work on your own, you might make a mistake and they could help you. Like, because, say, like, you can work in pairs because that isn't cheating, you can use a calculator, you can look at an answer and try working it out but if you're . . . Cheating's like if you think you know but you don't.

In Chapter 3, I described how Year 10 Ben worked hard on his noughts and crosses (Tic Tac Toe) investigation, but was strongly driven by his aim of completing more and finishing faster than the other students. Correspondingly, he states a preference for working alone. Ben talked enthusiastically about the set system and his competitiveness and belief in effort, the importance of speed, and tests. Since the final year in primary school he had determined to improve at mathematics, and be average as opposed to below average. He had revised for the SATs in both years 6 and 9 and welcomed the pressure from teachers to do well. He enjoyed mental arithmetic tests because they emphasized speed:

It's all the different stuff, it's pushing you, you're trying to get it, you're trying to get answers as quickly as you can.

He was similarly competitive about doing investigations, invoking "separate knowing" values of power and control, algorithms and completeness among others:

You want to find more, like say with the noughts and crosses you like doing the grids and you're wanting to find more than other people.

While "separate knower" lower-set boys projected aligned but positive identities, many girls in the sample, including those who we might classify as constructed knowers, described themselves as insecure in mathematics. The girls in the top sets appeared less confident and more vulnerable to the opinions and actions of others, showing more similarity with girls in lower sets in this regard. This was the case even for Year 10 Rachel, who, as we have seen in Chapter 3, had been promoted from a lower set at the end of the previous school year. She enjoyed investigations because they allowed space for generating her own understanding:

Because you're not just like being told do this, do this. You can like work out things for yourself and do it differently to how you normally do it.

Unlike the boys, however, Rachel tended to interpret her desire to understand and connect as weakness, and along with the other girls expressed a strong fear of failure. So, as we have seen in Chapter 3, Year 9 Rosie had strong dislikes for areas of mathematics that she found difficult, and Lizzie was relieved that she was no longer in the top set and the demands it made. Jenny, in the same set as Rosie, also described the top-set teaching as adversarial and test-driven, in a way which bears little resemblance to the teacher-as-resource picture painted by Michael.

While boys as well as girls could be classified as constructed knowers, these data show a sharp contrast between different performance groupings: it appears as I suggested in Chapter 3 that students in different sets develop significantly different epistemologies of mathematics. This difference is cross-cut by a gender difference in self-positioning, however: top set girls show a tendency to interpret a "constructed knower" outlook as vulnerability and failure in a way which is similar to that of the undergraduate women. The pattern that emerges here is complex, then, and it becomes more so when we look at the primary school data.

Finally, then, I return to Black's primary school data, drawing here on Black (2004b). As we have seen in Chapter 2, a major finding of this study is the difference in the quality of teacher–pupil talk when different groups of children are compared: middle-class boys tend to be involved in more "productive" exchanges with the teacher in the sense of dialogic interactions in which they are positioned and position themselves as genuine interlocutors in the learning project. There is also an indication that this basic distinction is cross-cut by ethnicity—as suggested by the cases of Hasan, who is Gujarati, and Janet, who is Chinese (see Black, 2002a for a fuller account). In what follows, we see that there is a further layer of explanation to add in the case of Sian, a middle-class girl who is described by Black as an "anomalous case" in that, with one exception, she is involved in far more exchanges than the other girls in the class. However, a close inspection of her participation in these exchanges shows that it is qualitatively different from that of the middle-class boys, and so she is anomalous in this sense too.

An important starting point in this analysis is the establishment of Sian as a high-attaining pupil, as shown in the following direction from the teacher to the whole class:

> T: Can you just remind yourselves when you come to multiplication that you don't do what Sian's been doing. So it just shows even the best of us can make mistakes. She's started with the tens and she's said "two sixes are twelve and the six nines are fifty-four" and adding it and starting with the tens when you don't, do you? You always start with the units. Six nines are fifty-four pennies and the five will go into the ten pences. So make sure you aren't making that mistake.
>
> (Black, 2004b, p. 8)

However, this does not mean that she is frequently engaged in the kind of talk that we saw in the case of Phillip in Chapter 2. Interactions which are typical of Sian are like the one which follows, in which she supplies a one-word correct answer at the end of an exchange involving boys. The task is to map nets onto solids:

T: Which one do you think they're meaning? Why are they calling it a squared prism do you think, not a rectangular prism? *(Phillip, Tim, Sean, and Erica put hands up)*

Tim: The mint creams.

T: Pardon?

Tim: The mint creams because they've got a erm square on the . . .

T: On the base and the top of it? I think they're referring to the Smartie box now . . . cos it's got a square base and a square top whereas a rectangular prism has got rectangular sides all the way around it. Out of those two the mint creams box is a rectangle and the Smartie box is the square isn't it? So I actually think that's what they're after. I think they're after the Smarties for that one. Now I don't want you to draw the net. I don't think you need to draw the net. All you need to do is tell me what it is. *(pause, Daniel puts hand up)* What is it? Can anyone tell me what that's gonna become?

Daniel: That? *(pointing to yellow net) (Tim puts hand up)*

T: No the green one. What kind of shape is that one? Sian? *(Ben puts hand up)*

Sian: A rectangular prism.

T: A rectangular prism. That's all you need to write for that. So for number seven just write rectangular prism.

(Black, 2004b, p. 9)

This is an example of conversational repair between the teacher and Tim which is similar to that involving Phillip in Chapter 2. The misunderstanding which is indicated by Tim's answer to the first rather than the second of the teacher's questions is cleared up and the dialogue continues with time for justification on both sides. In contrast, Black points out, when Sian is called upon she presents a minimally correct answer, which is accepted by the teacher without a request for justification. In the context of the whole discussion and repeated similar incidents, Black suggests that Sian is regularly used as a pace-maker in the class, a pupil who can be relied upon to provide a correct answer and so allow the lesson to move on, repairing time "lost" in discussions such as that involving Tim: she is positioned as a "*domestique*":

> I borrow the term *domestique* from the world of professional cycling where team members work to support one rider (usually the team leader) with the aim of placing them in an advantageous position in the race. If we apply this idea to our analysis of Sian's role in classroom discussions, another implicit outcome of her role is to allow space and time for the boys in Group A [the productive talk group] to engage in exploratory talk where they are able to talk themselves into understanding . . . (Black, 2004b, p. 10)

Sian's taking up of the position of domestique is not simply a female version of Group A membership: Black notes that this role is also assigned occasionally to boys in the group. However, these boys also participated in more prolonged exchanges, whereas this was not the case for Sian. Unlike them, she does not appear to have "the communicative rights to use teacher–pupil interactions as a tool for constructing and acquiring knowledge" (Black, 2004b, p. 17). A demonstration of these rights is evident in Chris' behaviour in the next extract, where the pupils are engaged in a subtraction task, but Chris has made an error:

Chris: Is that right?

T: What number's that? Twenty-two er twenty . . . no. That's wrong. Twenty-two, twenty-three . . . that's right and that's wrong. Er . . . one twenty-five that's wrong. Now what ya doin' wrong? Ah five from nothing is nothing is it? Gosh! *(teacher leaves the discussion)*

Nelson: Nothing take away five.

Chris: Five from . . . it says nothing take five.

Nelson: What?

Chris: It says nothing take five.

Nelson: I know. Nothing take five . . . take five.

Chris: You can't.

Nelson: Exactly.

Chris: Yeh that's what I've done.

Nelson: How come it's wrong?

Chris *(angrily)*: I don't know.

(a few minutes later, teacher re-enters the discussion)

Chris: How is this one wrong?

T: Because you can't take five from nothing.

Chris: Er . . . I haven't.

T: You are.

Chris: Nothing take five is nothing.

T: You've nothing. How can you take five from nothing?

Chris: You can't.

T: So what do you do? So that should be a ten then.

Chris: Oh yeh.

We can contrast Chris' persistence in getting the teacher to explain here with Sian's action in a similar situation, where she is calculating averages:

Sian: Mrs. Williams, I'm stuck on fourteen because I've added all those up and it comes to twenty but . . .

T: Share by five.

Sian: Five . . .

T: There's five numbers ain't it? If it's two numbers you should add them together and share by two. Is that what you did?

Sian: Oh right.

Sian and Chris act very differently in these two extracts: Sian does not question the teacher any further, but we might suggest that perhaps this is because she does not need to. However, Black tells us that in fact she did not understand, and had to seek help from Black herself. The teacher also acts differently in the two exchanges—she gives Chris "clues" in a way that could be seen as scaffolding, but simply tells Sian a partially complete method which contains a clue (division by 5 versus division by 2) but is not explicit enough for Sian to understand, although her "oh right" signals to the teacher that she has. Black speculates on the longer-term outcome of her role in other exchanges:

> Although it seems apparent that such peripherality would mean she is unable to take ownership of mathematical knowledge in the way that Group A do, her continuing capacity to be known as "one of the best of us" suggests her identity is not one of marginalisation. Perhaps her future trajectory may be towards an identity of compliance to the demands of compulsory schooling by performing well on academic tests but [not] an identity of full participation which can be transferred to the wider mathematics community and the academic capital associated with it. (Black, 2004b, p. 19)

This is just one story, of course, but it suggests a possible way in which mathematical lives develop so that girls can be at the same time successful in mathematics but feel that they are not. The joint production by Sian and her teacher of limited opportunities for building knowledge may be repeated for some time, and may feed and be fed by assumptions about girls' and boys' "ways of knowing." There is some support for this analysis in the data reported by Creese et al. (2004). Although it is not their intention to analyze the following extracts in terms of approaches to knowing (rather, they are concerned with the ways that boys and girls use language and talk about collaboration in the classroom), the data suggest that in fact middle-class boys (the data was drawn from two middle-class schools) are positioned as creative or constructed knowers, in Becker's typology, whereas girls are positioned as separate knowers. Here we see their teacher comparing boys and girls:

> I've always had this theory that men are quite often more interested in the process and females are more interested in the result. Now, I'd hate to say that, these are just generalities—I mean we are taught not to make these sweeping comments—and that's what these are in this case. So with girls you will notice they will just get on with the writing of it and not dwell too much on it. Whereas I will often find boys who get stuck on a sentence and they are not sure if they should do it like this and it's not right, and they sit down and have a chat (laughs). (Creese et al., 2004, p. 194)

Given these assumptions on the part of the teacher, it is perhaps unsurprising that the boys make the following comparisons, positioning themselves and the girls in the same framework:

Oscar: I reckon that girls learn differently, because what they do is, when they get a piece of work they . . .

Shaun: They seem to rush it.

Oscar: Yes, they sort of put it into their book, like paraphrasing it, so they have their own reference to a piece of information they've been given, but with boys they get that information and they sort of mould it, they sort of reshape it, so it's in different form, and they put different ideas into it . . . You could say that girls have a tendency to learn like people would probably learn in the late Victorian times, and boys learn like most people would learn now. (*interjections confirming this thesis*)

Oscar: I think there's two ways of learning really, there's sort of copying and getting information into your head . . .

Brendan: And getting it right.

Oscar: And getting it right. Yeah, and um . . . Staying within the perimeter, you could say. And then the other type is sort of learning, taking information, and then putting it down in a different form and then sort of piecing it together like a puzzle.

(Boys' friendship interview, Millbank) (Creese et al., 2004, pp. 194–5)

Changing Classroom Cultures, Not Changing Mathematics

The data that I have discussed here show that school mathematics is not as straightforwardly "pro-separate" as Becker suggests. While it has a surface emphasis on right answers, formality and certainty which is reflected in the discourses of natural ability and speed which circulate within the classroom community, it also supports a creativity and connection-making which some students access and incorporate into a positive mathematics identity. In Black's primary school data these students seem to be predominantly middle-class boys who participate creatively and are rewarded for doing so by further dialogue with the teacher which not only signifies their position in the content of their talk but also in its dialogic and intersubjective structure. The top-set boys in my secondary school sample described themselves as good at mathematics and similarly positioned themselves as engaged in an apprentice-type relationship with the teacher as epistemic resource rather than social authority. Top-set girls on the other hand appear to be caught between an institutional positioning as good at mathematics (i.e. they are in the top set) and a range of discursive positions which invoke gender differentiation in behavior and thinking and the particular performance marker of speed. At secondary school and university they may aim for creativity but find it difficult to sustain positive identities— their accounts are dominated by a fragility in terms of the constant danger of finding themselves out of their depth in terms of their own standards for understanding and performance and—crucially—others' recognition of that performance. Boys and men on the other hand sustain positive identities despite, for some, their arguably limited and limiting approach to mathematics.

Together with the findings of the research that I reviewed in the first section of this chapter, these patterns suggest that class, culture and gender are important organizers in students' developing relationships with mathematics. These relationships are fluid of course, but powerful gender discourses such as those voiced by Creese et al.'s teacher and pupils and the performance discourses that we saw in Chapters 2 to 5 constrain the range of identities that are available to mathematics learners. At the same time, institutional practices of audit and accountability serve to pressurize both teachers and students. Sometimes students resist these pressures and discourses, but not always successfully: compare how the students at Farnden and Middleton developed their own collaborative community of practice, and Debbie at Bradley self-authors as someone who has "a right to learn," while Janet, in Chapter 2, accepts the offered position of a "quiet girl" and Lizzie at Northdown School criticizes her new lower set for its lack of investigative mathematics but says she is happier there because of its less exposing demands.

As reform mathematics has indicated (NCTM, 2000), encouraging students to develop identities of participation in which they engage in the production and validation of mathematical knowledge, practice proofs for themselves, and appreciate the aims of mathematics and how it is used, makes mathematics accessible to many more students than otherwise in the sense that they are given the means of gaining ownership of it. Becker's own suggestions aim at dismantling teacher authority and finding "voice"—but this needs to be extended to include the development of an explicit mathematics literacy which will make mathematics accessible to all by connecting that voice to agreed processes of validation. She rightly calls for mathematics to be taught as a process:

> Mathematics knowledge is not a predetermined entity. It is created anew for each of us, and all students should experience this act of creation. Presenting mathematics . . . as disembodied knowledge that cannot be questioned, works against connected knowing. The imitation model of teaching, in which the impeccable reasoning of the professor as to "how a proof should be done" is presented . . . is not a particularly effective means of learning for women. (p. 168)

Nor, of course, is it an effective way of learning for men. As Schoenfeld (1994) points out, mathematics is "a living, breathing discipline in which truth (as much as we can know it) lives in part through the individual and collective judgments of members of the mathematical community" (p. 68). When it comes to owning mathematics and creating it for ourselves, it is our relationship to authority which is the crucial issue; as Raman shows in her analysis of proof procedures, doing mathematics involves both the key idea *and* the warrant, but where authority resides when we own mathematics is not in the teacher, it is in the mathematics itself, and this is central to Schoenfeld's case:

> Mathematical authority resides in the mathematics, which—once we learn how to heed it—can speak through each of us, and give us personal access to mathematical truth. In that way mathematics is a fundamentally human (and for some, aesthetic and pleasurable) activity. (p. 68)

So, finally:

> Learning to think mathematically means (a) developing a mathematical point of view—valuing the processes of mathematization and abstraction and having the predilection to apply them, and (b) developing competence with the tools of the trade, and using those tools in the service of the goal of understanding structure—mathematical sense-making. (p. 60)

The change that this suggests is one which recognizes the importance of undoing power relations in the social structure as a whole and which makes knowledge accessible to all by its demystification in a participatory pedagogy which encourages exploration within a framework of a mathematical literacy which makes its rules and the means of changing and developing them explicit. I explore these ideas further in Chapter 8.

8
Supporting Mathematical Literacy

In the previous chapter, I concluded that the major issue in exclusion from mathematics is the way in which its central practices are hidden from many students, causing them to remain on the margins, lacking the means of ownership. The multiple strands of individual learners' mathematics histories—their "trajectories through Discourse space"—combine to generate their unique relationships with mathematics. Patterns are visible, however, and we have seen how the discourses of gender, learning and of mathematics itself contribute to repeated positionings of self within a constrained range of available identities. My major focus in this book has been the extent to which these positionings of self enable identities of participation in mathematics, and I have shown that even students who strive to be creative participators frequently experience their relationships with the subject in terms of marginalization rather than participation. While gender discourses seem to support boys and men in more positive relationships with mathematics, this does not necessarily mean that they have identities of engagement. For students who are outside of the dominant white, middle-class culture, the practices by which mathematics is created and validated are simply not made available or visible in mathematics talk and texts. In the UK, these students are more likely to find themselves in lower sets, where pedagogic style again offers restricted options for developing a participatory identity. My over-arching concern, then, is with the means by which students can be supported to develop a mathematical literacy in the sense of participation in and ownership of the practices of mathematics.

In this chapter, I consider two related issues in supporting mathematical literacy. The first relates to the issue of the (in)visibility of practices and the need to make these explicit in order to empower learners. I will argue that we need an *inclusive* mathematics literacy which echoes critical literacy approaches in order to achieve this. The second acknowledges that the practices of mathematics are part of an established community: this means that mathematics is not therefore negotiable without participation in that community. I will argue that we need to think the teacher–learner relationship through in terms of power and authority in considering what entering a community of practice entails. I draw these two issues together in the final chapter which looks more closely at the implications for practice of an inclusive mathematics literacy, and the challenges which it inevitably presents.

A Question of Literacy—Making the Invisible Visible

In Chapter 1, I briefly reviewed some of the issues in the language of mathematics, beginning at the word level and moving on to the mathematics register. I return to this issue of text-level language in this section. There are two major reasons for doing so: the first is that, as we have seen in the previous chapter, there is evidence to suggest that the way in which language is used in mathematics texts makes a substantial difference to the way in which learners engage with it, and how they are positioned by it, as apprentice mathematicians or as practitioners of a mathematized numeracy event, to build on a distinction suggested by Barwell (2004). This is not a trivial observation for the reasons that Dowling supplies: we use language to make meaning, and different texts which appear superficially to be the same can in fact carry very different meanings. I shall suggest, as does Sfard (2000), that we need those meanings in order to do mathematics, and that we cannot replace them with something which is apparently more accessible without losing—or in fact obscuring—meaning. The second reason is that this observation indicates something of what we need to do in order to make meanings transparent and available to all learners.

One response to concerns about equity in mathematics education has been to challenge "dominant numeracies," in parallel to the rather more established challenges to dominant literacies exemplified in the work of the New Literacy Studies (see Street, 2003, for an overview). Such challenges have not been made without debate, however, and it is useful to inspect these debates in terms of their implications for mathematics before moving on to the case of mathematics itself. As I shall show, the central issues concern whether an emphasis on "voice" and authorship acts as a force for change or whether it neglects the role of genre in the shaping of meaning by language and is thereby merely "authorizing disadvantage" (Gilbert, 1994).

Insights from Literacy Pedagogy: Process and Product

A major development in recent literacy research has been the recognition that literacy develops before school (Williams, 1998), continues to develop throughout schooling (Christie, 1998), and in addition develops *outside* of schooling (Barton & Hamilton, 1998). It has also been recognized as an essentially social phenomenon (Barton, Hamilton, & Ivanic, 2000), and, correspondingly, as a rapidly changing one in an age of techno-literacy (Lankshear & Knobel, 1998; Snyder & Lankshear, 2000). These issues have had an impact on pedagogic theory and practice in various ways. One such is the view that schooling should build on the non-schooled, everyday, literacy practices which abound in our society. These may be literacy practices which are displayed by the pre-schooler (Hall, 1994), or they may be those which form part of our daily adult lives (Barton & Padmore, 1994); these latter may also be culturally-specific literacies such as traditions of story-telling (Kenyon &

Kenyon, 1996) or constructions of writing and reading as social and shared activities (Heath, 1983). These practices can theoretically be absorbed into pedagogy: everyday and non-standard literacies should be not only recognized and valued, but can be used as a bridge into school literacy. Some educators argue furthermore that these everyday literacies should form the basis for a substantial challenge to what is seen as the dominant literacy (Baker, Clay, & Fox, 1996; Esterhuysen, 1996; Gallego & Hollingsworth, 2000). What this means in practice is that, because it is a basic and early written (and oral) form, there is an emphasis on *narrative*, not only in the writing classroom, but as a possible vehicle for a number of school subjects, particularly science (Esterhuysen, 1996; Heath, 1996; Kenyon & Kenyon, 1996). I return to this issue below with respect to mathematics.

Another way in which the development of our understanding of literacy has influenced pedagogic theory and practice has involved a focus on the dominant literacies themselves—the literacies of schooling. As Christie (1998) has emphasized, literacy learning is a continual process that spans the years of schooling, with "significant milestones . . . if students are to emerge with a reasonable degree of proficiency . . . There is a relationship between the changing patterns of written language that students need to recognize and use, and the variety of knowledge they need to learn" (pp. 49–50). It is this latter link that is crucial, and there are parallels to draw with mathematics as I shall show. Using Systemic Functional Linguistics (Halliday, 1994), Christie maps the development of writing from early "and then" recount, which is based on personal experience and is closer to speech than to "mature" writing, through a middle years explanatory or procedural mode which shows command of tense, technical and explanatory language and an authoritative author role, to later years expository or argument-based writing, which is marked by control of grammatical metaphor and the removal of agency. A major feature of this development is the ability to control these genres. Given that communities differ in their relationships to text (Heath, 1983), it is not surprising that the successful development of control over high-status genres and academic registers is related to social and cultural background (see Martin, Christie, & Rothery, 1987; Schleppegrell, 2004). It is this fact that leads to some major issues in pedagogy.

When it comes to the teaching of writing, this is usefully described in terms of two possible (although not necessarily incompatible) positions: an emphasis on process, authorship and creativity on the one hand, and an emphasis on control of genre and text-level form on the other. The recognition of more personal and culturally-based literacies frequently coincides with a pedagogic shift towards a "progressive", "process writing" pedagogy; it stresses authenticity and creativity and a shift of power away from the teacher as assessor and sole audience to the child as author writing for a variety of audiences. Children are taught to go through a series of stages such as pre-write, draft, conference, revise, edit, publish, with a crucial emphasis on authorial voice and lack of interference or judgment from the teacher (Czerniewska, 1992). But as a backlash against traditional approaches which focused on the rules of "correct" (i.e. dominant, standard) English expression and their embodiment in the final product to the neglect of content, understanding, creativity and form, the process writing emphasis on "finding voice" risks losing sight of the importance of the product itself, leaving teachers with no theoretical basis for directing children's writing. The emphasis on authoring, creativity and personal expression has two outcomes: it favors a bias towards recount or narrative in school writing (Gilbert, 1994), and it directs attention away from the way in which a text is constructed and why (Martin et al., 1994). A significant consequence of this latter is that the teacher has no grounds for intervening in any direct way; she has no rationale for evaluation or critique of a piece of writing. In addition, learners are not given the opportunity to develop command of the higher-status genres—as Schleppegrell (2004) points out:

> Without a focus on form and attention to different register expectations, process-oriented approaches can easily become trivialized. Students may not be pushed to attempt unfamiliar genres if they are only encouraged to be expressive and write on whatever topics they wish. This deprives students of opportunities to learn more than they already know and means that students are not prepared for writing in all subject areas. (p. 149)

This is not to say that we should neglect process and context altogether however, but that process and product are complementary. Furthermore, the critique of the emphasis on authoring highlights an important equity issue: although we may all be able to recognize and identify different genres, we may not be able to construct a text to fit the bill without considerable guidance and scaffolding, and access to cultural resources. Thus Christie (1998), Martin, Christie, and Rothery (1994) and Schleppegrell (2006, 2007) among others, argue for the explicit teaching of the specific features of academic genres, since those children who are not given the opportunity to develop a confident command of written genres and an awareness of their functions and variation are disadvantaged. Teaching them explicitly about genre will liberate them from the ties of recount and narrative which are the natural genres for accounts of personal experience: even when children have different audiences and purposes for writing, narrative dominates because teachers frequently do not know how to explicitly support other genres, and yet "every teacher is a language teacher, responsible for helping all students develop the linguistic tools that will enable them to learn and share what they have learned" (Schleppegrell, 2006, p. 11). The argument from the point of view of functional linguistics is that the bias towards narrative is damaging, limiting experience of the powerful meaning-making expository and analytic genres; unless teachers are in a position to make reference to such linguistic structures in assessing pupils' work, they will not be able to make written language available to *all* students. What distinguishes this approach from earlier pedagogies which emphasized form is that it is a *critical literacy*—that is, by making the structure of mainstream academic texts explicit in terms of the way in which knowledge is constructed, it empowers learners to engage with and even challenge that knowledge (Macken-Horarik, 1998). As Christie (1998, p. 70) argues, "Much conventional wisdom about literacy learning never really acknowledges the very considerable developmental process that is involved in learning literacy, nor does it acknowledge the important role of teachers in understanding the features of writing, and in teaching students to be aware of these features, in order to become effective users of literacy themselves."

Functional linguistics presses a strong case against the process-writing classroom, then. Martin et al. argue that process writing offers not choice but *pseudo-choice*, condemning the under-privileged to imprisonment in a fantasy world (if that). As the discussion so far suggests, one of the fundamental differences between these two positions is what each assumes about what kind of knowledge or skills we are dealing with when we are talking about writing, and what is involved in acquiring such knowledge. While both take the view that literacy is a political, cultural, and social issue, they differ on the point of entry of the individual into the process. Thus genre theorists emphasize the wider social traditions of which the individual becomes a part, while those who would challenge dominant literacies emphasize processes of negotiation between individuals in the construction of knowledge. These are important differences, and they determine to some considerable degree what the role of the teacher should be. These distinctions are even more apparent when we transfer the discussion to the development of mathematics literacy.

Parallels in Mathematical Literacy

Challenges to dominant literacies which claim that narrative can operate as a suitable and more accessible vehicle for subjects which traditionally use exposition or argument necessarily assume that scientific and mathematical knowledge can be properly conveyed within narrative. A number of writers have argued that this is indeed the case with respect to science (Esterhuysen, 1996; Heath, 1996; Kenyon & Kenyon, 1996). In the area of mathematics it is most clearly articulated by Burton (1995, 1996, 1999a), whose argument for a narrative approach has its basis in her stance on mathematics as a socio-cultural artifact, with consequences for pedagogy and the development of under-standing:

> I claim that a narrative approach to mathematics and its pedagogy is consistent with a view of mathematics as being socially derived and with the understanding of mathematics as being socially negotiable. ... By

engaging with the narrative, we place the mathematics in its context and personalise it, making it come alive to the conditions of the time. Context provides meaning . . . By narrating, we make use of our power to employ language to speculate about, enquire into, or interrogate that information. (Burton, 1996, pp. 32–3)

There are parallels here with Becker's position in terms of an emphasis on the personal, although while Becker's difference feminism leads to a focus on the pedagogy of mathematics, Burton's more radical feminism leads to an epistemological challenge, as she shows in Burton (1995). In this particular piece, she argues that the public face of mathematics as established and non-negotiable knowledge leads to an invisibility of its actual processes of development, along rather the same lines as discussed in the previous chapter. Thus she quotes Polanyi and Prosch (1975, p. 63) as arguing that "the processes of knowing . . . are rooted in personal acts of tacit integration. . . . Scientific enquiry is accordingly a dynamic exercise of the imagination and is rooted in commitments and beliefs about the nature of things." Burton takes this argument one step further to argue that we need "a theory of . . . knowing, as subjectively con-textualized and within which meaning is negotiated" (Burton, 1995, p. 211), and it is this basic stance that underpins her suggestion that we see mathematics "as a particular form of story about the world . . . Re-telling mathematics, both in terms of context and person-ness, would consequently demystify and therefore seem to offer opportunities for greater inclusivity" (Burton, 1995, pp. 213–4).

Similar arguments to those rehearsed in Chapter 7 apply here. Recognition of the role of intuitive insight does not mean that mathematics is negotiable on the level of individuals: it *is* negotiable on the level of the community of practice, however, in terms of debate about how claims to knowledge can be justified and defended, and thereby what is in fact known. The intrinsically social nature of mathematics means that we need to operate not on the level of individuals but on that of the community, and so the inclusivity that Burton is rightly concerned with is a question of how individuals gain access to the meaning-making practices that are central to the community rather than of how an individual voice can be heard. Indeed, she recognizes the role of the com-munity when she says: "I am claiming that knowing, in mathematics, cannot be differentiated from the knower even though the knowns ultimately become public property and subject to public interrogation within the mathematical community" (Burton, 1995, p. 220). Thus an emphasis on authorship (Burton, 1999a; Povey, Burton, Angier, & Boylan, 1999) suggests a similar problem to that raised in the debate about literacy: it is not clear how an emphasis on narra-tive and the personal enables real access to mathematics in terms of the practices of mathematical reasoning and justification that enable agency within the community. If we take this seriously, however, an essential element of such a "process mathematics" is conflict between different perspectives and multiple strategies, reflection on these, and the consequent generation of deeper understanding. However, Burton's (1996) example of two different narratives describing resolution of the familiar "crossing the river" problem does not really capture this process, and in fact one of the narratives does not seem to actually address the problem itself. It goes like this:

> First we got cuisennaire cubes and got out two cubes for the boys and two cubes for the men and made a model of it all. We started moving the cubes about. It looked easy at first but we soon found it wasn't so easy. We were moving the cubes about for some time and then we had got it into a position that we saw we were almost there. We saw the answer and moved the cubes and they were all on the other side. (p. 34; corrected for spelling)

In contrast, Burton quotes the following rather more mathematically helpful text which is accompanied by cartoon drawings, each portion of text appearing in a different cartoon frame numbered 1 to 12:

> Two boys go over/ 1 boy comes back/ 1 man goes over/ 1 boy comes back/ 1 man goes over/ 1 man comes back/ 2 boys go over/ 1 boy comes back/ 1 man goes over/ 1 boy comes back/ 2 boys go over/ they're all there (p. 35; corrected for spelling)

These answers to the problem are very different. Text 1 is written in a personal style and the past tense, in narrative fashion (it even has a complication—"we soon found it wasn't so easy"), and it describes the circumstances *around* the solution to the problem, although it doesn't describe the problem solution itself. As mathematics it is, essentially, vague; although it is interesting, it doesn't address the problem in terms of presenting a solution or even part of one. It might have some function as a description of insight in problem solving ("we saw we were almost there. We saw the answer") but it does not tell the reader what the answer is or how to arrive at it herself. In contrast, Text 2 combines words and pictures to tell the reader how to solve the problem; it is impersonal, it is written in report style and it uses simple present tense and short sentences. It does not use capitals and it uses numerals rather than written number words; it also employs a "key" to symbols. Indeed, the cartoon pictures stand alone as a solution without the written text at all. To this extent the solution uses symbol more reliably than text (which is not self-explanatory alone, in fact, and in this regard is inadequate). These two texts are fundamentally different; they are not only written in very different styles, but they are about *different subject matters*. Only one of them is mathematics.

The issue being raised here is one of how far it is actually possible to express mathematics within a personal narrative without some sort of serious loss in meaning or purpose. Clearly, Burton would like this to be possible: the idea of using narrative in mathematics teaching is an attractive one because it appears to present a solution to the inequities in mathematics that are a major concern for many mathematics educators. Like many writers, she is also concerned with the idea of fostering connectedness in mathematics teaching. Her project is to enable an epistemological change in which "Knowing mathematics would . . . be a function of who is claiming to know, related to which community, how that knowing is presented, what explanations are given for how that knowing was achieved, and the connections demonstrated between it and other knowings . . ." (Burton, 1995, p. 221). Although this functions well as a pedagogical change, it is questionable as an epistemological change. However, the examples above do not seem to support the claim for narrative as a central part of making either change happen—indeed, they suggest the reverse. But to really examine the possibility of a narrative approach to the learning of mathematics, we need to look beyond these two examples: we need to distinguish between less important practical considerations such as the tendency when employing narrative to gloss details, omit them or even fabricate them (see White's (1987) critique of the use of narrative style in science) and the more fundamental issue of the possibility of developing and expressing knowledge of a particular kind in an unusual genre. Are the more familiar genres through which mathematics is expressed simply a manifestation of a political, social, or institutional bias, or are they integral to the meaning-making of mathematics? I will address this question in two ways: firstly by looking at an authentic example of mathematics taken from history, and secondly by returning to the contribution of functional linguistics in an analysis of how mathematics texts make meaning.

Mathematics and Narrative in Historical Perspective

An examination of the history of mathematical texts and the development of the modern formal style might at first sight offer support for the view that mathematics can be expressed in narrative. First impressions can be misleading, however. The kinds of text that are taken to be the norm for the modern mathematician are the end result of an increased "standardization" of texts and a narrowing of linguistic options that developed throughout the nineteenth and twentieth centuries. In the nineteenth century, a large number of different forms were used in mathematical texts. In particular, since the journal paper was in its infancy, the narrative style more typical of the older forms of communication, in particular of the letter, was widely employed by mathematicians in order to present their results (for a discussion, see O'Neill, 1993). Letters themselves often made their way into publication and in these texts mathematics is embedded in a narrative. Consider for example the following extract from William Rowan Hamilton's presentation of his discovery of quaternions:

> MY DEAR GRAVES,—[1] A very curious train of mathematical speculation occurred to me yesterday, which I cannot but hope will prove

of interest to you. [2] You know that I have long wished, and I believe that you have felt the same desire, to possess a Theory of Triplets, analogous to my published Theory of Couplets, and also to Mr Warren's geometrical representation of imaginary quantities. [3] Now I think that I discovered yesterday a *theory of quaternions* which includes such a theory of *triplets*.

[4] My train of thought was of this kind. [5] Since $\sqrt{-1}$ is in a certain well-known sense a line perpendicular to the line 1, it seemed natural that there should be some other imaginary to express a line perpendicular to both the former; [6] and because the rotation from 1 to this also being doubled conducts to -1, it also ought to be a square root of negative unity, though not to be confounded with the former. [7] Calling the old root, as the Germans often do, i, and the new one j, I inquired what laws ought to be assumed for multiplying together $a + ib + jc$ and $x + iy + jz$. [8] It was natural to assume the product

$$= ax - by - cz + i(ay + bx) + j(az + cx) + ij(bz + cy);$$

[9] but what are we to do with ij? [10] Shall it be of the form $\alpha + \beta i + \gamma j$? [11] Its square would seem to be $= 1$, because $i^2 = j^2 = -1$; and this might tempt us to take $ij = 1$, or $ji = -1$; [12] but with neither assumption shall we have the sum of the squares of the coefficients of 1, i, and j in the product $=$ to the product of the corresponding sums of the squares in the factors. [13] Take the simplest case of a product, namely the case where it becomes a square; [14] we have $a^2 - b^2 - c^2 + 2iab + 2jac + 2ijbc$ and $(a^2 - b^2 - c^2)^2 + 2(ab)^2 + 2(ac)^2 = (a^2 + b^2 + c^2)^2$; [15] the condition respecting the *moduli* is fulfilled, if we suppress the term involving ij altogether, and what is more, $a^2 - b^2 - c^2$, $2ab$, $2ac$ are precisely the coordinates of the *square point*, so to speak, deduced from the point a, b, c, in space, by a slight extension of Mr Warren's rule for points in a plane. (Hamilton, 1844, pp. 106–7)

The style of the letter might seem to substantiate the arguments of those who defend the arbitrariness of the modern mathematical paper and the possibility of expressing mathematics through narrative. We have here a narrative, indeed a personal narrative, in which mathematics is presented.[3] This narrative presentation is found again in Hamilton's notebook accounts of quaternions:

16th October 1843

(1) I, this morning, was led to what seems to me a theory of quaternions, which may have interesting developments. (2) Couples being supposed known, and known to be representable by points in a plane, so that $\sqrt{-1}$ is perpendicular to 1, it is natural to conceive that there may be another sort of $\sqrt{-1}$, perpendicular to the plane itself. (3) Let this new imaginary

be j; (4) so that $j^2 = -1$, as well as $i^2 = -1$. (5) A point x, y, z in space may suggest the triplet $x + iy + jz$. (6) The square of this triplet is on the one hand $x^2 - y^2 - z^2 + 2ixy + 2jxz + 2ijyz$; (7) such at least it seemed to me at first, because I assumed $ij = ji$. (8) On the other hand, if this is to represent the third proportional to 1, 0, 0 and x, y, z, considered as *indicators of lines*, (namely the lines which end in the points having these coordinates, while they begin at the origin) and if this third proportional be supposed to have its length a third proportional to 1 and $x^2 + y^2 = z^2$, and its distance twice as far angularly removed from $1, 0, 0$ as x, y, z; then its real part ought to be $x^2 - y^2 - z^2$ and its two imaginary parts ought to have for coefficients $2xy$ and $2xz$; (9) thus the term $2ijyz$ appeared de trop, and I was led to assume at first $ij = 0$. (10) However I saw that this difficulty would be removed by supposing $ji = -ij$. (Hamilton, 1843, p. 103)

Both of these narratives contrast with Hamilton's presentations of quaternions in his papers submitted to journals. His account of quaternions for the *Philosophical Magazine* begins thus:

1. Let an expression of the form

$$Q = w + ix + jy = kz$$

be called a *quaternion*, when w, x, y, z, which we shall call the four *constituents* of the quaternion Q, denote any real quantities, positive or negative or null, but i, j, k are symbols of three imaginary quantities, which we shall call *imaginary units*, and shall suppose to be unconnected by a linear relation with each other. (Hamilton, 1844–50, p. 227)

The shift between the formal and impersonal style of the mathematical paper and the narrative style of the notebook and letter appear to be a matter of choice determined by the social and institutional context and their associated different literacy practices. The letter and the notebook belong to the informal narrative genres of everyday practices, the mathematical paper to a different, formalized context in which narrative is expunged. It is in virtue of the shifting nature of the academic institutions in the early part of the nineteenth century that the narrative style could survive in a public and visible form as a letter published in a journal. Later developments have rendered such possibilities unacceptable. The conclusion might then be drawn that mathematical knowledge can be expressed within a variety of different genres, and would be consistent with Burton's view that there is nothing about mathematics that rules out the employment of a narrative genre.

However, while it is true that there has been a closing of the textual forms that are permissible in formal academic institutions, this conclusion is too hasty.

An examination of the letter and the notebook reveals a more complex structure than a simple narrative. The texts contain two distinct component texts: a *mathematical* text is embedded within a *personal narrative*. The difference between the texts is indicated in the tense system, the use of deictic reference and the forms of lexical cohesion employed.

Consider first the letter. The narrative text proceeds for the most part in the past tense. Introduced in the opening paragraph [1]–[3], which twice employs the deictic reference "yesterday," the narrative proceeds from [4] onwards in simple narrative structure in the past tense in which the "train of thought" is recounted: [5] "it seemed natural"; [7] "I inquired"; [8] "it was natural to assume" and so on. The cohesion in the text is achieved through that narrative order. This text needs to be distinguished, however, from the mathematical subtext from [5] onwards. The sentences and sub-clauses in which the mathematical equations themselves are presented use verbs which are tenseless: the indicative sentences employ the "timeless" present. Consider [5] and [6]: "[5] Since $\sqrt{-1}$ *is in a certain well-known sense a line perpendicular to the line 1*, it seemed natural that *there should be some other imaginary to express a line perpendicular to both the former;* [6] *and because the rotation from 1 to this also being doubled conducts to –1, it also ought to be a square root of negative unity, though not to be confounded with the former.*" While the narrative super-text indicated in the "it seemed natural" remains tensed, the mathematical sub-text, in italics, is in the timeless present. At the same time there is also in the mathematical sub-text a distinct form of cohesion: the temporal order in the narrative gives way to a logical order in the mathematics: [5] since . . . there should be; [6] because . . . it ought to be . . .

Similar points are true of the notebook entry. We are presented with a narrative sequence of past tense verbs that refer to a temporally ordered sequence of events. This temporal sequence is further fixed by the employment of temporal conjuncts, (7) "at first," and deictic reference (1) this morning. As in the letter, the mathematical sub-text shifts into the timeless present. Hence the shift in (2) to "1 is perpendicular" and the timeless reading of the "=" signs in the equations in (4), (7), (8), (9), and (10). The cohesion of the narrative is realized through a temporal order—sequences of superordinate clauses with past tense verbs, temporal conjuncts and temporal deictic reference—whereas the cohesion of the mathematics is realised by a logical order: e.g. "so that" in (2) and (4) and the complex if . . . then conditional of (8).

These features of the mathematical text which distinguish it from the surrounding narrative are neither accidental nor arbitrary: they are constitutive features of mathematical discourse. Even if it is embedded in a narrative structure, mathematical argument is itself atemporal—one cannot say for example, "yesterday $\sqrt{-1}$ was in a certain well-known sense a line perpendicular to the line 1, but today it no longer is"—and it achieves cohesion through logical rather than temporal order. Mathematics is constituted by a logical structure that is not reducible to a temporal sequence of events, and we need to distinguish between the variety of different kinds of genre in which mathematics can be *embedded* and the textual structure of mathematics itself. Mathematics can be embedded in a variety of texts in a variety of styles from dialogue (consider Plato's *Meno*) and narrative through to the modern impersonal text of the journal paper with its house style and narrow linguistic options. There is room for criticism of the closing of options in modern mathematical papers and the forms of exclusion this might foster. This, however, is quite distinct from linguistic features constitutive of mathematical discourse itself: mathematics cannot be narrative for it is structured around logical and not temporal relations. In this example, while the mathematics is embedded within a personal narrative, a focus on the authorship as signalled by that narrative obscures the very real role played by the author's (in this case Hamilton's) control of the mathematics discourse which, in Schoenfeld's (1994) sense "speaks through all who have learned to employ it properly" (p. 62).

Mathematics and Genre

One of the potential complications of the analysis of literacy from a functional linguistics perspective is that it has simply replaced one set of rules with another, and has become equally prescriptive as a result, re-imprisoning those who it set out to liberate—this criticism is raised by Barrs (1994), for example. The illustration from Hamilton's work points to a partial truth in this claim. There do exist restricted genre types which are the product of specific institutional patterns of professionalization. However, not all meaning is separable from the generic form in which it is construed. As Martin et al. (1994) point out, the inseparability of meaning and form has fundamental implications for teaching:

> The common sense duality of meaning and form is not a harmless one. In education in particular it is pernicious and has proved a major stumbling block for progressive initiatives. The whole movement toward child-centred education has foundered on the idea that children can understand and undertake history, geography and other subject areas "in their own words". That this is a necessary starting point, no one would deny . . . But that children should be stranded there . . . is impossible to accept. It cuts them off absolutely from any real understanding of what the humanities, social sciences and sciences are on about and denies them the tools these disciplines have developed to understand the world. (p. 237)

Insofar as genre shapes and constrains the nature of a text, then graphs, equations, proofs and algorithms can be considered as expressions of genre; as Marks and Mousley (1990) argue, mathematics utilizes a number of genres:

In solving problems, writing reports, explaining theorems and carrying out other mathematical tasks, we use a variety of genres ... Events are recounted (narrative genre), methods described (procedural genre), the nature of individual things and classes of things explicated (description and report genres), judgements outlined (explanatory genre), and arguments developed (expository genre). (p. 119)

They note, however, that few of these genres are evident in mathematics classrooms, and that where explanation appears as the province of the teacher, not the pupil, and where it is teachers who draw logical connections and construct proofs rather than enabling pupils to do so, they will be excluded:

Let us not make the process writing mistake again. We should avoid encouraging the use of students' language only within the limited and limiting models of language used in many creative writing sessions, and work actively to change the linguistic models and expectations currently presented in many mathematics classrooms and resource materials. If we are to set out to teach students to write different genres in mathematics, the role of the teacher is more than just to observe, prompt and react to students' efforts to express their mathematical ideas verbally, in writing or in other forms. It is also to teach the variety of genres valued in adult communication. (Marks & Mousley, 1990, p. 132)

This teaching needs to be explicit, however, if learners are to gain control of the genre. Thus, while Povey, Burton, Angier, and Boylan (1999) emphasize voice and authorship, they underplay the issue of how learners can gain access to the tools of critique and evidencing which they also acknowledge to be a central part of an inclusive classroom:

In mathematics classrooms in which learners are the author/ity of knowledge, they have the opportunity to use their personal authority both to produce and to critique meanings, to practice caring in a dialogic setting where the effectiveness of their own narrative(s) and also those of others is refined. ... When the learner's understandings do not fit with those of others, they are encouraged to engage ... in the practice of critique, a practice fundamental to creating potentially emancipatory discourse. (Povey et al., 1999, p. 237)

Just as research by Martin et al. (1987) with aboriginal children found that left to their own devices they produced no more than recount, so students given the task of writing about mathematical processes but without clear direction from the teacher as to how they should do this produced vague accounts which focused on outcome rather than process (Fortescue, 1994, cited in Schleppegrell,

2007, p. 152). Only when the teacher had modelled the task were they able to develop the facility to write down explanations and procedures. Similarly, Shield and Galbraith (1998) found that encouraging students to write about mathematics had the effect of developing an authorial voice, but not their mathematical thinking: "the students simply became more skilled at presenting their algorithmic style of writing" (p. 44), reproducing the textbook style they were familiar with. Shield and Galbraith conclude that, without a change in teaching practices, students cannot develop more meaningful engagement with mathematics. As Lampert (1990, p. 41) suggests, the role of the teacher in a classroom where the aim is for students to develop a "new way of knowing mathematics" is to "make explicit the knowledge she is using to carry on an argument with them about the legitimacy or usefulness of a solution strategy . . . She is modelling an approach to problem solving." Authorship and agency need to be founded on a control of the means by which meaning is made in mathematics and a consequent self-positioning within the discipline:

> An analysis of their assertions shows that they are not mere expressions of opinion; they recount their own reasoning processes and analyze those of others. They use "I think" to mean "I have figured out that," and their assertions of what they have figured out are regularly followed by arguments for why their strategies seem valid. They are indicating authorship of the ideas that they assert, and they are also indicating that thinking is something that a student both does and talks about in a mathematics class. Making assertions in this form is an expression of what they believe about roles and responsibilities in relation to mathematical knowledge and where they put themselves in relation to the establishment of valid arguments in the discipline. . . . Their choice of refutation as a form of disagreement with their peers . . . is a significant indication of what they believe about how truth is established in mathematics: it is *not* established by the teacher, or another student, saying that an answer is right or wrong, but by mustering the evidence to support or disprove an assertion . . . (pp. 54–5)

All this is not to say, however, that narrative in mathematics is not of use, simply that it serves just one function among the many needed. It may provide important context, as part of a study of history of mathematics which leads to a critical perception of maths in context as indicated by Povey, Elliott, and Lingard (2001). It may also serve a function as a personal history, in the sense of access to one's own reasoning, which could, in theory at least, then form part of practicing explanation and justification of mathematical claims. However, as critical literacy theorists have argued, and as I have argued in the previous chapter, we need to take care not to relegate certain learners to a narrative or purely personal approach. For example, O'Halloran (2003) presents findings which are similar to Dowling's textual analysis and Bartholomew's work on "ability" groups and which suggest that teachers use less formal and technical language when teaching working-class and female students. Her analysis suggests that this fact restricts access to mathematics:

> *At the expense of the mathematical content,* the linguistic selections in the oral discourse of the lessons with working class and elite private school female students appear to be more consistently oriented towards interpersonal meaning. Furthermore, in these lessons the tendency towards a deferential position in power relations and an overall semantic orientation do not accord with that found in mathematical discourse ... The greater incidence of non-technical mathematical register items and consequent lack of taxonomic relations, together with the non-generic board texts, suggests that shifts or scaffolds to mathematical discourse do not occur in the lesson with working class students. (O'Halloran, 2003, p. 210, my emphasis)

What is crucial is that teachers need an awareness of mathematics as a discourse as it is played out at several levels from word to text and the ways in which learners respond to this. As Lampert's analysis shows, enabling learners to develop an identity of participation requires "some telling, some showing and some doing it with them along with regular rehearsals" (p. 58), and hence a teacher role of expert and guide. I explore the implications of this below.

Entering the Community of Practice: Power, Authority and Emotion

As my own data and those of other researchers show, students experience exclusion from mathematics for a number of reasons: there are patterns in pedagogic practices which favor some but not others, while students themselves bring particular histories, experiences and beliefs with them to the classroom which inscribe the identities and positionings with relation to mathematics which are available to them. Beliefs about mathematics are often enacted within classroom discourse and practice, itself driven in many respects by institutional constraints and government-level demands. Of prime importance among these beliefs are those which emphasize speed and fixed ability; in a climate of high-stakes assessment these simply serve to re-emphasize and reinforce practices which differentiate between learners.

Within this already divisive context, the impact of the complexity and density of mathematical language on learning and the demands that it makes is considerable. Class and culture constrain the availability of identities of participation from the outset for some learners, but the extent of the gap between everyday numeracy and mathematical discourse creates further barriers. There is much that is linguistically invisible, and traditional teaching, aided by powerful discourses about the nature of mathematics, compounds this invisibility by obscuring some of the central practices of mathematics. There is also evidence to suggest that it is selective about who it reveals certain practices to, and that white middle-class boys hold an advantage in comparison to other class members in terms of their self-positioning within this dynamic. It is this complex interweaving of factors which generates our relationships to mathematics that an inclusive mathematics pedagogy must address by enabling a refiguring of identities and the development of agency. Such a pedagogy needs to:

- make mathematics language explicit and underline the fact that mathematics is a community of practice with highly specialized forms of communication;
- give pupils opportunities to take possession of ways of making meanings, and to talk themselves into understanding;
- reveal the central practices of mathematics by laying bare the structure of mathematical argument and encouraging learners to try this for themselves;
- emphasize participation so that the identity of "mathematician" is available for all learners, who are given the means to take up this position;
- emphasize how learners can take possession of rules in the practice and become rule-makers themselves.

There are, therefore, very specific roles for teachers and the nature of the teacher–pupil relationship is crucial to an inclusive mathematics pedagogy. I address these issues here, with particular attention to power, authority, and emotion, encapsulated in the following vignette from Bibby (2006):

Researcher: What is the best way of learning in maths lessons?
(*Italicized comments refer to Steven, their usual teacher*)
Pupil: *Doing it off the board and the teacher helps you out* [be]cause when you do it from the books she [the student teacher] just speaks and gets annoyed.
Pupil: *If you get to speak in front of the whole class explaining why you think it's this and like everyone sharing their own methods*, instead of the teacher we've got right now is a trainee and she says we can only use this method. She only gives us two methods and we have to use them. *Otherwise we can all share our own methods* and I think it would be more easier.
Pupil: Sometimes we ask her like for help and she doesn't understand it. She tells us what she thinks I said and then walks away and then helps like five other people before she comes back.
Pupil: We need some expert teacher, that's what we need.
Researcher: So you learn best when you feel you're being taught by someone who's got very good knowledge?
Pupil: Yeah.
Pupil: Like Mr Blake [Steven].

Pupil: Yeah, everyone learns a lot.

Pupil: Yeah, it's a bit funny. Sometimes we're naughty like when some children choose to be but I think we learn better when we know someone's actually an expert and will actually listen to you.

Pupil: He can handle the class well.

Pupil: If people play up he will deal with them.

(Key Stage 3 discussion, age 12–14; Bibby, 2006)[4]

The central issue raised by a concern with developing agency in mathematics for all learners is that of power relationships and their adverse effect on learning for some students in particular. At the same time, acknowledging that the practices of mathematics are part of an established community means that mathematics is not therefore negotiable without participation in that community. I have argued, therefore, that equality can only be secured by explicit awareness of the language, function and usage of mathematics and an appreciation of the nature of mathematical knowledge claims. This places the teacher in a position of expert, a situation which educationalists working in many subject areas but particularly in mathematics, perhaps, are anxious to avoid because of its potential to deny learners access to understanding. Many would see the students' positioning of Mr. Blake as an expert and their account of the student teacher's actions in the extract above as equally worrying. There is a difference, however, in how Mr. Blake exercises his expertise and how the student teacher exercises hers. In this section, I will argue for a rather more positive but also a more fine-grained account of the expert role and also for its inevitability. I will propose that, not only does the nature of mathematical knowledge mean that teaching and learning *will* be done by experts and novices respectively (and, following Christie (2003, p. 178), I suggest that it is dishonest to claim otherwise), but that learning in the first place relies on the epistemological authority of the teacher-expert and the learner's willingness to take on trust a mathematical way of doing things. Their "legitimate peripheral participation" (Lave & Wenger, 1991) will ultimately enable an articulated participation in the practice, but this takes time. The distinction between *epistemological* authority and *social* authority is key to this process; it is the effect of social authority to which Edwards and Mercer (1987) (and the students in the extract above) rightly object.

One of the criticisms of standard teaching practices is that they support merely "ritual" as opposed to "principled" knowledge (Edwards & Mercer, 1987), syntax but not semantics (Hull, 1985) or, similarly, semantically-debased or pseudo-structural conceptions rather than meaningful constructions (Cobb & Yackel, 1993). Using the appropriation metaphor, Ernest (1991) makes a comparable distinction between knowledge owned by the teacher and knowledge owned by the pupil: while the ideal might be knowledge owned by the pupil, the reality frequently is ownership by the teacher. Thus Vygotskian and neo-Vygotskian approaches to the classroom construction of knowledge such as Edwards and Mercer's (1987) and Mercer's later work (1994, 1995) maintain that the shaping of "principled" knowledge relies on the continuous production of shared mental contexts or frames of reference culminating in a handover of competence from guide to apprentice (cf. Rogoff, 1990). Context, by this definition, is a property of the understanding of a situation which is jointly negotiated or constructed by the participants. For Edwards and Mercer and many others, "principled" knowledge is something to be aimed at as "quality" knowledge versus empty syntactic rule-following. Their most illuminating example comes from an example of a group of secondary school students' failure, in a lesson on pendulums, to grasp, among other issues in scientific method, that of controlling variables (i.e. allowing only one at a time to vary) and their acquisition instead of knowledge which is embedded in the trappings of the lesson ("You couldn't have us all doing the same thing, so we/there was three of us and there was really three things to change on the pendulum so we done one each"). We can say, then, that a knower has "principled" knowledge when she shows power to project from what has been learned in the past to appropriate and correct behavior in a variety of new and previously unencountered situations: in a new situation, she knows how to proceed.

There is a potential confusion in the idea of ritual and principled knowledge in that it suggests that ritual knowledge is not useful—it is mere recipe-following. In mathematics, rule-following without understanding is something that most people have experienced, and as I showed in Chapters 4 and 5, this is sometimes acceptable to learners and sometimes not. In Wenger's terms, it equates to alignment, a mode of belonging that is not necessarily a negative one, although it can certainly be so. We can for instance say that coming to participate in a community of practice *requires* a stage of "legitimate peripheral participation," a stage at which the learner is a potential participant but is not a fully skilled practitioner. This is easier to see if we consider principled knowledge not as "knowing that" but rather as a question of "knowing how," and although knowing how *can* be a question of merely copying or habitually reproducing behavior or blindly following instructions in certain circumstances, it can also be a more situationally embedded knowing about how to proceed within a social practice. Participation, even at the periphery of the practice, is part of learning:

> A community of practice is a set of relations among persons, activity, and world, over time and in relation with other tangential and over-lapping communities of practice. A community of practice is an intrinsic condition for the existence of knowledge . . . participation . . . is an epistemological principle of learning. (Lave & Wenger, 1991, p. 98)

However, a view of knowing as being a participant in a community of practice and of knowledge as intrinsically social has important consequences for how we come to know: because it is the practice which constitutes the meaning, it cannot be the case that, for instance, we apprehend mathematical "truth" and then learn how to describe it according to current convention. Clearly, teaching mathematics cannot be a mere demonstration of "the mathematics all around us," because mathematical meaning is constituted by practice—we do not see the evidence and then see mathematical truths. As Perelman (1963) points out, the teacher cannot teach by pointing to the evidence alone, although in many ways this is precisely how mathematics is presented, as I showed in Chapter 5:

> It is the *notion of evidence* as characterising reason that must be challenged ... Evidence is thought of as the force before which every normal mind must yield and at the same time a sign of the truth of whatever imposes itself because it is evident. Evidence would bind the psychological to the logical, and allow passage from one of these levels to the other. (Perelman, 1963, pp. 136–7)

What this suggests is that the teacher's task cannot be to point out what already exists in the real world, but to induct children into talking mathematically about it. The only way in which she can do this is to practice such talk herself and for learners to participate in those practices with her—they do not learn *from* talk, they learn *to* talk as I argued in Chapter 1. As Schleppegrell (2007) notes, "students develop mathematics concepts as they use them discursively to construe meaning" (p. 148). Thus Walkerdine's (1988) analysis (pp. 122ff) of a nursery school addition lesson shows that what happens is far from an exercise in pointing out to the children the mathematics all round them. In her description of this episode she shows how, in her highly repetitive talk and action, the teacher encourages the children to participate in a discourse which describes groups of blocks and the actions performed on them in a particular, *mathematical*, way, moving from the physical action and its everyday description ("put them together") to use of the mathematics register ("three and four make . . .") to, ultimately, the written mathematical form ($3 + 4 = . . .$). She provides a way of talking which is appropriate to her conception of the action, the conception she wants the children to have. So, as Ryle (1949) argues, and again this is very pertinent to how mathematics is portrayed in teaching:

> When teaching *how* or when learning *how?* I think it is clear, without much more argument, that didactic expositions of arguments with their conclusions and their premises, of abstract ideas, of equations, etc., belong to the stage after arrival and not to any of the stages of travelling thither. (p. 280)

Early participation may then consist of a certain amount of "joining-in" behavior which, although it may be at another's bidding and in that sense not autonomous, is nevertheless situated within the discourse—"apprentice learners know that there is a field for the mature practice of what they are learning to do" (Lave & Wenger, 1991, p. 110). As Sfard and Lavie (2005) comment in relation to early learning: "the process of becoming a participant of the numerical discourse is inherently circular: To become aware of this discourse's advantages one has to use it; yet, to have an incentive to use it, one has to be aware of the prospective gains of this use. Following in the footsteps of more experienced interlocutors is probably the children's only option" (p. 288).

As we have already seen, teachers' use of context to support children's learning can only succeed when that context is mutually interpreted and understood. In establishing such a joint frame of reference, however, it is necessarily the teacher's understanding of the situation which dominates: her task is to get the children to join in with and agree to *her* interpretation. But if she is not to slip into the production of ritual knowledge, in Edwards and Mercer's sense, this agreement must not be founded merely on the teacher's social power to demand *token* agreement. The learner for their part must take at least part of what a teacher says on trust, such trust being founded on recognition of the teacher's status as a mature practitioner. In so far as a learner cannot do the "right" thing alone (as Wood, Bruner, & Ross' (1976) definition of scaffolding suggests), they do not truly share a mental context with the teacher. Thus a learner must experience a period of not understanding, but of going along with the activity in hand nevertheless, chiming in the right words in the slots provided by the teacher. This activity will be based on the learner's perception that there is a *right* way of doing things and, as I have argued elsewhere (Solomon, 1989), the appreciation of the authority of "significant others" (cf. Hamlyn, 1978; Sfard, 2006; Sfard & Lavie, 2005). Teachers have epistemological authority: as Augustine (1977) remarks, "with regard to the acquiring of knowledge, we are of necessity led in a twofold manner: by authority and by reason. In point of time, authority is first; in order of reality, reason is prior." What a teacher in fact does is encourage pupils to join in with particular ways of talking about the things they are doing; the context is shared because the teacher, by means of "tuning in" to the learner's current under-standing of the situation and adjusting what she says accordingly, ensures that it is. It is not so much a question of the teacher's use of the "tricks of the trade" which Edwards and Mercer identify as supportive of ritual learning, but of her engagement in a dialogue with the learner which thus ensures the crucial shared mental context ("doing mathematics") which may more accurately be called the *product* of learning rather than (or as well as) its prerequisite. The background to this process is a mutual attempt to understand—Sfard and Lavie (2005) provide an example in relation to mother–child counting episodes: "the needs of grown-ups and of children complement each other and both these sets of needs stem from the underlying common need for communicating" (p. 292). This need is also present in formal teaching and learning situations.

However, as Adler (1997, p. 255) points out, there is a tension between the need to make mathematics language and culture visible, and the need to guide students towards mathematical understandings:

> While the *withdrawal* of the teacher as continual intermediary and reference point for pupils enables [the teacher's] participatory classroom culture, her *mediation* is essential to improving the substance of communication about mathematics and the development of scientific concepts. That is, both are required, and managing the tension is the challenge!

Distinguishing epistemic authority from social authority is one way to reconcile the apparent contradiction between the need to make mathematics visible and the description of learning as in some part at least ritual. As Sfard (2006, p. 168) says, "Students' persistent participation in mathematical talk when this kind of communication is for them but a discourse-for-others seems to be an inevitable stage in learning mathematics. . . . To turn the discourse-for-others into a discourse-for-oneself, the student must explore other people's reasons for engaging in this discourse." I have suggested that the teacher–pupil relationship relies on trust, but this does not necessarily entail blind or unquestioning belief in a relationship in which the teacher simply talks and pupils simply listen. A trust relationship needs to go beyond this to one in which learners can question and challenge teachers. Burton (1999a, pp. 23–4) tells a story about students' disbelief and anger on finding that a published mathematical paper contained a serious error. An inclusive pedagogy relies on dismantling the kind of power relationships which engender such a reaction. So in addition to making the mathematics transparent, such a pedagogy also has to make the social relationships of the classroom visible, in ways which enable students to ask questions and to recognize that teachers are not always right but that this does not at the same time make them wrong. As Becker (1995) argues, students need to see lecturer fallibility—but that will only be a useful experience in a wider context of a teaching and learning community which explicitly distinguishes social and sociomathematical norms (Franke, Kazemi, & Battey, 2007, pp. 238–9; Lampert, 2001; Yackel & Cobb, 1996).

So to return to the extract at the beginning of this section: Mr. Blake is not just an expert mathematician, he is also an expert teacher—and it may be that when the students say "expert" in this extract they do in fact mean the latter—it is not clear. In any event, the important distinction is how he uses his expertise to guide but not to constrain—his authority lies in his ability to demonstrate the practices of mathematics, rather than in knowing the "right answer": he encourages other explanations, and he listens. The student teacher does neither of these things. As someone in a position of power, she appears to abuse the students' trust. She does not let them suggest alternative strategies, she does not listen, she quashes their enthusiasm and refuses to help. Bibby (2006; forthcoming) reports that learners aged 9–11 and 12–14 appear unable to talk about mathematics without reference to their teachers, demonstrating the intensely emotional nature of their relationships. We see this also in the following extract, this time from a Key Stage 2 (ages 9–11) discussion:

P1: OK, Miss South, she sometimes ignores me and stuff and I don't really like it, yeah she sometimes ignores it, I don't really like it [. . .] It hurts my feelings.

Researcher: It hurts your feelings, yeah, and I wonder does that make it easier to learn or harder?

P1: Harder, and like I open my book and I see that I get all the questions wrong, wrong, wrong, wrong.

Researcher: Oh no, so you feel ignored and then you're feeling that you haven't got it right? [. . .]

P1: . . . and it's not my fault that I got everything wrong, it's Miss South's fault that I got everything wrong.

Researcher: So what happens when you get it wrong?

P1: I feel guilty.

Researcher: You feel guilty?

P1: Yeah.

Researcher: Because?

P1: Miss South ignored me [. . .]

P1: I don't know why I feel guilty . . . it's still quite my fault because I got the questions wrong a bit, but it's normally Miss South's fault.

P2: Well I feel like that, it's because Miss South like, I put my hand up and she never chooses me, especially like in maths, she loves the other Year 5 class and then like um she blames me if I've got it wrong, it's like "Hafsah you don't understand" but it's her, she doesn't understand and then when I'm ignored I don't like it, I feel left out and nobody ignores me! But then she says I only ignore you, it's because you're so clever, but then that's not true.

Researcher: What do you think is true?

P2: I think it's just she doesn't like me, no Alice it's true, I don't think she likes me that much.

P2: When you keep on being left out, you think of something, you think that oh maybe they don't like you and that's how *I* feel.

Researcher: And what happens to you when you're trying to get on with learning things and you feel that your teacher doesn't like you? Does that make a difference?

P2: Yes, it's a bit difficult to like concentrate and then she's like *"you're not concentrating properly"*, but then when you tell her that *"you're leaving me out"* then she doesn't know how you feel because it's not happening to her!

Researcher: Because she's not being left out?
P2: Yeah cos she's being, like everyone's surrounding her going Miss South, Miss South!

In these extracts we see the emotional investment of learners in their relationships with teachers, and their hurt at betrayal of trust. This is not just a social issue, it is a pedagogic one: their complaints about the student teacher and Miss South are based in the effect of the relationship on their learning. Classroom relationships are at the heart of relationships with/in mathematics.

9
Relationships with/in Mathematics
Inclusive Mathematics Literacy in Practice

In their review and theoretical framing of recent equity and diversity research, Cobb and Hodge (2002) set out a number of issues which resonate with my overall project in this book in explaining how mathematical identities develop. My primary aim has been to understand how different individuals in the same classroom can develop very different relationships with mathematics, by capturing the complexity of the ways in which they are positioned within mathematics communities and at the same time position themselves within their ongoing narratives or "history-in-person" —"the sediment from past experiences upon which one improvises, using the cultural resources available, in response to the subject positions afforded one in the present" (Holland et al., 1998, p. 18). Similarly, Cobb and Hodge are concerned to explain how:

> ... the gatekeeping role that mathematics plays in students' access to educational and economic opportunities is not limited to differences in the ways of knowing associated with participation in the practices of different communities. Instead, it also includes difficulties that students experience in reconciling their views of themselves and who they want to become with the identities that they are invited to construct in the mathematics classroom. (p. 249)

This complexity has implications for solutions to the problem of exclusion—a recognition of the role of student agency in the process of developing mathematical identities means that this also needs to be taken into account in understanding students' responses to the way that mathematics is taught. Thus developing classrooms which theoretically enable better access to mathematics is only one side of the story—promoting inclusion also depends on students opting to, or having the ability to, take that access up and to maintain their sense of who they are—or, as I have suggested may be the case for some students, to refigure their sense of self with respect to mathematics. While groups of students who engage in home and local community practices which are valorized within mathematics classrooms may be motivated to accept access, those whose discourse practices do not fit with classroom expectations may choose to resist the opportunities which are offered because they do not fit in identity terms. Thus, in their discussion of Boaler's work (Boaler, 2007) and Gutiérrez's work (Gutiérrez, Baquedano-López, Alvarez, & Chiu, 1999), Cobb and Hodge observe that their particular recommendations "involve changing the practices of the mathematics classroom to make it possible for students to reconcile the identities that they are invited to construct with types of identities that they value" (p. 277). In this final chapter, I consider the challenges of an inclusive mathematics literacy with these issues in mind.

Accessing Mathematics Discourses

As Morgan's (2005) critique of the DfES (2000) guidelines for teaching mathematics vocabulary shows, improving access to mathematics is not merely a matter of teaching vocabulary, but of enabling students to engage in mathematical discourse, including mathematical argument. This is a complicated aim, however—as Moschkovich (2002) points out, clarifying what we mean by mathematical discourse and what counts as competency in mathematical communication or even what counts as mathematics is itself a challenge. Sfard (2000) addresses this concern in her discussion of the interpretation of the NCTM (1991) *Standards* emphasis on discourse: because the very nature of mathematical discourse as practiced by professional mathematicians is not necessarily translatable into a more grounded classroom discourse, we have a dilemma of how to proceed, if we want to recognize students' need to understand. Acknowledging that the *Standards* do not necessarily imply that schools should attempt to replicate professional mathematical discourse—"One may rightly expect that classroom mathematical discourse would be a greatly relaxed, less rigorous, and more 'popular' version of this discourse" (p. 180)—she cautions against total compromise, for reasons similar to those rehearsed in Chapter 8:

> A compromise, however, can count as reasonable only as long as the discourse we are left with is well defined, intrinsically coherent, and generally convincing. If, on the other hand, we simply reject some of the basic conventions without replacing them with alternative rules, or if the changes we make are accidental and inconsistent, we may end up with a discourse so ill-defined and amorphous that it simply cannot be learned. (p. 183)

So how can students be given access to the discourse? Like Lampert, Sfard argues that the process of making the discourse explicit must include teachers as expert practitioners, because "rules of language games can only be learned by actually playing the game with experienced players" (Sfard, 2000, p. 185). Indeed, Schleppegrell (2007, p. 148) cites MacGregor's (2002) finding that "students working in groups are not always able to express their ideas clearly or understand each other's explanations," while Adler's (1997) study of participatory-enquiry lessons suggests that this approach "can inadvertently constrain mediation of mathematical activity and access to mathematical knowledge" (p. 235) because learners are unable to explain their understandings: "*He can't hear her question and she can't hear his explanation*" (p. 248). Similarly, Chapman's (1995) study of intertextuality in school mathematics shows that without strategies for connecting texts students have difficulty in communicating with each other in their explanations. Unless they have teacher support as illustrated in Forman, Larreamendy-Joerns, Stein, and Brown's (1998) work on collective argumentation, for example, students working together are less likely to use mathematical language. Veel (1999) reports that they are more likely to use everyday language, failing to exploit the meaning-making potential of the kind of grammatical resources illustrated in Chapters 1 and 8: for example, the lexical density of their talk is only one third of that in teacher talk, and there are more short nominal groups, as opposed to long ones in teacher talk. As Lubienski (2007) observes, there is an important distinction to be made between gaining from discussion of conflicting viewpoints as a general pedagogical principle versus its exercise as an essential part of doing mathematics: "if it is an important end in itself, then efforts should be made to both help students understand the norms and roles assumed by such an approach, as well as to adapt the approach to students' needs and strengths" (p. 21). To this end, we can turn to Brenner's (1994) proposed communication framework, designed to systematically address the various elements of learning and doing mathematics. The framework comprises communicating *about* mathematics (reflection on thinking processes, communicating points of view), *in* mathematics (attention to register and representations), and *with* mathematics (meaningful problem-solving and the use of mathematics to assess arguments and their alternatives). I use this as an organizer here.

Communicating about Mathematics

A central tenet of reform mathematics is the promotion of student agency through discussion and argumentation. For example, Maher's (Maher, 2005; Martino & Maher, 1999) account of her 17-year study tracking the mathematical development of a cohort of students from their early schooling through to university shows how students provided with the opportunity and the tools for justification in mathematics in a collaborative learning environment are able to see themselves as participants in the practice. She describes how, given various problems such as the pizza problem (how many different pizzas are possible with *n* different toppings?), students worked together to produce answers and attempt to persuade other groups of their answers. Central issues are the use of argument, and also the making of public statements:

> The notion of argument was key to the way the students worked together. For example, to represent an idea, an individual may create a structure or

present a notation. In this way, the ideas are made public in their discourse in the form of explanations, actions, writings, and notations. . . . ideas presented earlier would again be visited, discussed, and thought about. This process was entirely student driven and made possible by individual initiatives. (Maher, 2005, p. 9)

It is important to note, however, that the students were not in a position of simply coming to mutual agreement about the justifiability of their claims: Maher tells us that they were required to "present their ideas with suitable justifications that were convincing to them (*and to us*)" (p. 2, my emphasis). They were frequently "challenged," to find out why rules work, to re-express their work in different notation, to re-assess it and to defend their claims as one of the students, now at university and looking back on the experience, explains:

You didn't come in and say this is what we were learning today and this is how you're going to figure out the problem. We were figuring out how we were going to figure out the problem . . . what would be our own formulas because we didn't know that other people had done them before. We were just kind of doing our own thing, trying to come up with an answer that was legitimate and that *no matter how you tried to attack it,* we could still answer it. It was a solid formula that works no matter how you tried to do it. (p. 9, my emphasis)

Another student makes the researcher demands even clearer:

If we tried to just present a final thing and really didn't know it from the beginning we couldn't explain it in a way that that you would accept from us. So in order to explain it in a way that you would accept we'd really have to start from bare bones, from the beginning . . . (p. 11)

In the UK, NRICH (Back & Pratt, 2007; Piggott, 2004; Pratt & Kelly, 2007; Smith, 2006) provides an "enrichment program" initially aimed at higher attainers from GCSE to undergraduate/graduate level, but moving towards the development of materials for all ranges of attainment (Piggott, 2004). The major focus of the project is to enhance mathematical problem-solving within group-learning workshops and whole-class discussion focused on argument and justification. Evaluation of the project (Smith, 2006) suggests that students' ability and willingness to explain their thinking is enhanced, in addition to their confidence with algebra. NRICH also appears to have encouraged persistence in the face of challenge, the ability to unpack problems and develop alternative solutions and independence from the teacher; where students are high attainers with low social status, project involvement improves confidence. Back and Pratt's (2007) analysis of the project's on-line learning environment addresses in particular the issue of how discussion boards support the development of an identity of mathematician. Their tracking of one Year 10 student's posting on the site suggests that he moves from tentative peripheral participation in his requests for help to a self-positioning as one who can answer other posts with "teacherly" help and advice, developing an identity which straddles the two communities of research mathematics and school mathematics.

Both of these projects report success but, as I described in Chapter 7, learners take up particular positions within the classroom which means they do not necessarily benefit from teaching which is designed to be inclusive. Thus Lubienski (2007) notes that NCTM *Standards* call for a particular classroom environment which in fact could end up simply privileging in new ways those students possessing the cultural capital to engage in the required way: an emphasis on discussion and exploration is not necessarily the kind of classroom culture that all students can straightforwardly access. She found that lower socio-economic class students talked about their role in discussions as that of obtaining or giving right answers, a response which seemed to be related to their perceived ability to make sense of discussions: "more lower SES students consistently said that having a variety of ideas proposed in discussions confused them. In general the confusion centered around feeling unable to discern which of the various ideas proposed in a discussion were sensible" (p. 16). In contrast, Lubienski's higher SES students described the value of listening to everybody's opinions and using these to arrive at a better answer. They were also adept in unpacking the teacher discourse of hinting that answers were incorrect rather than directly saying so in terms of how students were expected to respond, showing a "pedagogic awareness" in Black's sense. Thus she says: "I began to wonder if the way I inserted information into the discussions might have been more helpful to the higher-SES students, who seemed more attuned to my conception of, and rationale for, our roles" (p. 17).

As Brenner (1994, p. 244) points out, talking about one's thinking processes is not universal, and may appear alien to learners from some cultures or educational traditions. Thus encouraging students to communicate about mathematics is a particular focus of the QUASAR project (Silver, Smith, & Nelson, 1995; Stein, Smith, Henningsen, & Silver, 2000) which targeted economically disadvantaged students. In particular, QUASAR schools aimed to teach challenging mathematics to middle-school students at all levels of achievement, underpinned by classroom environments emphasizing cooperation and collaboration, and also trust. Silver et al. (1995) argue that the historical positioning of this group of students as non-participants means that this latter quality of QUASAR schools was crucial to promoting "cognitive payoffs when students are able to use their minds freely in exploring and exchanging their mathematical ideas" (p. 35). Enhanced relevance of the curriculum and explicit connection to students' lives and interests was an important feature for the same reason. Similarly, in Project Impact (White, 2003), teachers and students unpacked word problems in terms of their structure and aims, consciously drawing on real-world knowledge to discuss the nature of the problems and how they might be interpreted. Students were also encouraged to assess their answers in terms of their reasonableness—clearly a problem for many learners, as I showed in Chapter 7—and to reflect on their thinking processes and solution strategies rather than getting right answers. An important factor in this project was the fact that the students—predominantly African American and Hispanic—were unused to this kind of process; in line with other researchers, White observes that the norm is for teachers to assume a more controlling and monologic role with this group of students, but:

> As a result of asking children to share their thinking every day, children became more fluent and able express their ideas. That only happens when mathematical discussions become a classroom norm that is negotiated and changed. (p. 51)

Communicating in Mathematics

The second element of Brenner's framework includes attention to the complexity of the mathematics register and discourse, and how mathematical ideas are represented. As I have argued in Chapters 7 and 8, control of mathematics as a genre in terms of an explicit focus on its language is an essential part of an inclusive mathematics literacy. In an example of direct focus on mathematics language, Barwell's (2005) analysis of a classroom discussion of the ambiguity of the language of dimensionality initially introduced by the teacher but taken up by a student ("There's no such thing as a one dimensional shape coz a line is kind of like a rectangle filled in" (p. 123)) illustrates how greater understanding ultimately results:

> K's statement about one dimensional shapes ... and the ensuing discussion, can be seen as an exploration of what it is possible to say using the word dimension. In short, the students are probing and developing aspects of mathematical discourse. In so doing, the use of the term dimension becomes more complex, encompassing a range of mathematical discursive practices. ... the class engaged in a process of joint meaning-making, trying out the possibilities of words which they have encountered before extending their experience of using these words to think mathematically together. Ambiguity can therefore be seen as a learning opportunity rather than a hindrance. (p. 124)

As Barwell notes, the teacher takes an "unofficial" line in her initial introduction of the ambiguity of shapes which challenges the DfES (2000) guidance regarding the need to "sort out" ambiguities which students may have (rather than introduce them). However, in doing so, she opens up a discussion in which the students can engage in flaunting the official labeling of plastic shapes designed to support the activity, and in doing so underline how things are "meant to be." This process involves a complex cross-over or code-switching between formal and informal language which ultimately serves to clarify and enable mathematical sense-making. Similarly, Moschkovich (2007b) suggests that seeing students' contributions in discussion as inadequate because they draw on informal, everyday discourse as opposed to formal mathematical discourse misses the point that the overall shape of discussion may be mathematical in terms of its goals and focus. Thus in her analysis of third-grade bilingual students' discussion of the properties of a trapezoid, she argues that, despite their use of everyday rather than academic mathematical terminology, they are constructing shared meanings and in this sense doing mathematics:

> The definition students were using can be described as a working or stipulative definition. Using this perspective of definitions, we can see that these students were participating in an activity that may, in fact, be closer to the practice of scientists and mathematicians than to conventional school practices. (p. 28)

Reflection on this level demands particular knowledge on the part of teachers: Hill, Schilling, and Ball's (2004) analysis of teachers' mathematical knowledge suggests that teachers themselves need explicit rather than implicit knowledge of particular aspects of mathematics: they found that teacher knowledge was multidimensional, including "why mathematics statements are true, how to represent mathematical ideas in multiple ways, what is involved in an appropriate definition of a term or a concept, and methods for appraising and evaluating mathematical methods, representations, or solutions" (p. 27). These considerations are exemplified in D. L. Ball's (1993) reflections on her experiences of teaching reform-based elementary school mathematics to a culturally and linguistically diverse group of students with varying achievement levels. She demonstrates the familiar tensions involved in the concept of mathematics teaching as focused on learning to think mathematically—while the teacher's role is not one of pure telling, it is also the case that while they engage with whatever challenge the teacher has presented, learners need to be guided so that they arrive at a mathematics which is connected to the community at large rather than an idiosyncratic invention of their immediate group:

> Students must learn mathematical language and ideas that are currently accepted. They must develop a sense for mathematical questions and activity. They must also learn how to reason mathematically . . . Because mathematical knowledge is socially constructed and validated, sense-making is both individual and consensual . . . Thus community is a crucial

part of making connections between mathematics and pedagogical practice. (p. 376)

One of Ball's dilemmas famously involves the aim that "Mathematics teachers must respect students' thinking even as they strive to enculturate students into the discourse of mathematics" (p. 384): what to do with non-standard ideas and insights? As Ball (p. 385) points out, understanding what lies behind an unconventional claim is not easy—teachers do not necessarily share the learner's frame of reference and may not even see it. In the case of "Sean numbers," Ball did however understand the basis of Sean's claim, and did decide to legitimize his idea that numbers could be both even and odd, allowing her students to pursue a variety of investigations of "Sean numbers" despite her concern that they may become confused. In a follow-up quiz, the children were however able to correctly identify odd and even, and they did not invoke Sean numbers—in this instance, enabling Sean to experience "doing mathematics" in a fundamental sense despite his unconventional claims seems not to have done any harm. However, Ball's third dilemma concerns the issue of authority—she aims to avoid her own teacher authority "as the final arbiter of truth," but to "develop and distribute in the group a set of shared notions about what makes something true or reasonable" (p. 388). Again, the dilemma is about when to intervene, and how, how to deal with claims that are true in the immediate context but not in a wider, future one that the students will meet, and how to ensure that they *do* gain from the discussion, as opposed to becoming simply confused. Like Schoenfeld's (1994) problem-solving classes with much older students, Ball's students appear to have learned a self-reliance within the group to deal with their confusion—thus after a particularly demanding and unresolved session on negative numbers, Mei says "I'm going to listen more to the discussion and find out" (p. 392). While the danger of misconceptions being given as much time for discussion as correct ideas and thus gaining a hold constitutes a serious part of the dilemma, Ball argues that:

> . . . time spent unpacking ideas is time valuably spent. I have too often been confronted with evidence of what students fail to understand and fail to learn from teaching that strives to fill them efficiently with rules and tools. It is not clear to me that *telling* them that $6 + (-6) = 0$ will result in more enduring or resilient understanding, or in better outcomes in terms of what children *believe* they are capable of. (p. 393)

Rooting the mathematics register in real-world experience with the aim of empowering African American students, the Algebra Project (Davis, West, Greeno, Gresalfi, & Martin, 2007; Moses & Cobb, 2002) specifically focuses on moving students from representations of their everyday lived experiences in their own words and in pictorial form, to representations in everyday language, then to mathematical language, before finally moving to symbolic form. This strategy aims to provide students with "evidence for mathematics," based as it is in their own observations and natural language. The literal basis in experience (e.g. riding the subway) is designed not only to provide such evidence but also to maintain interest among a student group who do not historically identify with mathematics. Of particular interest in the current context is the issue of transparency, in the sense of explicating mathematical significance rather than algorithmic rule-use; Davis et al. underline the importance of this for re-positioning African American students as legitimate peripheral participants, in their analysis of an episode in which Moses discusses place value:

> In these interactions, Moses and his students constructed the students' positioning with significant competence and entitlement for under-standing the mathematical concept of numerical representation. The students were positioned as competent to understand how the repre-sentational system works, not just to follow procedures for manipulating symbols . . . (pp. 85–6)

It is worth noting that in this episode Moses is described as "quite directive," in terms of the shape of the discussion, its information content, and his evaluation of the students' responses. As I have argued in Chapter 8, his role as expert places him in a particular position of trust with his students; thus Davis et al. go on to say that:

> Along with this authority, Moses was accountable to the students for presenting the mathematical analysis that was needed. He was also accountable for reaching mutual understating with the students of the way that referential significance of the symbols is composed of the meaning of their constituents. (p. 86)

Communicating with Mathematics

Communicating *with* mathematics is the third of Brenner's framework elements. She makes the important distinction between students' recognition that achievement in mathematics is often key to access to further education or high-status careers, versus an understanding of the applicability of the mathematics they are learning to their current and future life. As she points out, merely seeing mathematics as a necessary "ticket" to the future may be a long-term goal that does not help students to sustain effort in what they may experience as meaningless but necessary evil, although Martin (2007, p. 157) argues that the African American adults in his study were empowered by their recognition of their exclusion in the past: "the same struggles typifying much of African Americans' experience can be transformative in terms of one's identity when those struggles are put in the context of liberation and freedom

. . ." In her study of literacy practices in underprivileged groups, Heath (1983) suggests that the benefits of connecting mathematics to the students' lived world via a focus on financial transactions, confusions and miscommunications at their local store enabled them to "understand the latching of details to the main idea, juxtaposition of facts and opinions, and the necessary sequence for incorporating critical details, numerical and otherwise, in word problems or any other recounting of an event" (p. 338). The critical element in these learners' experiences of keeping journals and analyzing the role of mathematics in the world is reflected in work by Frankenstein (1995), who also attempts to alert students not just to the utility of mathematics but also its role in the maintenance of social conditions, including its *mis*use in perpetuating inequity. Similarly, a major element of Gutstein's approach (Gutstein, 2006, 2007; Gutstein, Lipman, Hernandez, & de los Reyes, 1997) is developing tools for critical mathematical thinking and the use of mathematics for critical thinking (e.g. the cost of a B-2 bomber versus the cost of educating a student for four years). This "teaching for social justice" has a particular aim which is relevant here, in not only promoting academic success and raising political consciousness, but also developing agency and positive identities. Although, like Ball, Gutstein admits to a number of dilemmas—a critical mathematics is in danger of engendering relativism; there is a tension between teachers raising powerful issues but at the same time enabling students to raise their own; and ultimately there are limits to the use of mathematics to answer sociohistorical questions—he does report success in enabling students to gain the agency to "read the world with mathematics." Here he quotes Paulina, who still does not like mathematics, but says:

> Yes, I think I'm able to understand the world with math. All the math problems, projects, discussions about drug testing, Chicano history, etc., have made me understand because knowing about those issues and the discussions that we did made me think of what math might be involved. The math that we did helped me even more. (Gutstein, 2007, p. 64)

Classroom Relationships

A common theme running through the research on access to mathematical literacy is that communicating on all levels is dependent on the nature of classroom relationships and students' positioning within them: they need to be motivated and enabled to reflect and argue, to make suggestions and challenges, and to draw on their own experiences. Identity is key—to have access to mathematical reasoning, students need to see themselves as learners with particular ways of "doing school" in which all classroom members regardless of their perceived ability play a part in what is in effect a "counterculture" given conventional assumptions (Lampert, 2001, p. 65). However, as Horn's (2007) work, which I discussed in Chapter 2, and indeed the observations and interviews with Mrs. Williams, indicate, teachers are also positioned within particular discourses and communities, and perceptions of students in terms of "fast, slow and lazy kids" are often perpetuated through broader curricular and institutional practices that make a shift in perspective difficult, even when teachers are committed to change. Focusing on the implications of Wenger's theory for teacher identity within multiple communities, and the frequently observed mismatch between teachers' reform-based beliefs and their actual practice, Van Zoest and Bohl (2005) suggest that "understanding of the modes of participation—particularly relative to the classroom and school communities—. . . would allow teachers to act upon their knowledge and beliefs effectively in communities not necessarily supportive of them" (p. 340). They thus argue that the tensions for teachers are less a matter of internal inconsistency, but one of being able to reflect on their position within multiple communities and maintain their sense of themselves as effective teachers.

A possible insight into these processes and the tensions within them is suggested by Gutiérrez, Rymes, and Larson's (1995) concept of "third space," which has particular relevance in situations where learners are positioned, or position themselves, outside of, or resistant to, the mainstream classroom discourse. In the classroom dialogue which they describe, the teacher's assumption of what counts as knowledge (knowing what is in the *Los Angeles Times*) is carried through a series of closed "quiz questions" which the students cannot answer (they do not read this paper). While some students do participate in his monologic script, others begin to give voice to a "counterscript," in which they tell about local knowledge and make jokes which feed off of the teacher's script but are not part of it. While the two scripts carry on in parallel for the most part, they occasionally meet, in what Gutiérrez et al. call the "third space"—it is here that student and teacher cultural interests cross over and where there is potential for dialogicity. In the particular classroom exchange that is the focus of this analysis, both students and teacher back off from this uncertain territory. However, Gutiérrez et al. argue that, when a teacher appropriates a student's contribution and integrates it into the lesson, then this creation of a third space allows a negotiation of the lesson content which is not mere "bolt-on" multiculturalism:

> The construction of such classrooms requires more than simply "adding-on" the student script; it requires jointly constructing a new sociocultural terrain in the classroom where both student and teacher not only actively resist the monologic transcendent script, but, more importantly, also create a meaningful context for learning. (p. 21)

The crucial role of different classroom cultures in the success of reform curricula is underlined by Boaler (2006, 2008) and Boaler and Staples (2008), who discuss the promotion of relational classrooms (see also Franke et al., 2007,

p. 242; Schoenfeld & Kilpatrick, 2008) in an important challenge to the deeply embedded UK practice of "ability grouping" which, as I have argued, is a major contributor to identities of exclusion. They report gains in equity in terms of how the relational values of Railside school in California, focusing on respect for others and their viewpoints, created a "third space" which promoted positive learner identities resulting in students who "learned more, enjoyed mathematics more and progressed to higher mathematics levels" (Boaler & Staples, 2008, p. 609). In Railside mathematics classes, students focused on solving problems in heterogeneous groups, with particular emphasis on responsibility for each other's learning. In these "multidimensional classrooms" multiple methods and solutions were valued, as were different ways of approaching the task as a whole. Given this multidimensionality, it is not surprising that an emphasis on justification was found to be crucial in the success of Railside school. What is particularly significant however is the way in which this related to the promotion of equity; Boaler and Staples (2008) report that the wide range of achievement, basic starting knowledge and motivation in the classes, coupled with the goal of responsibility for each other's learning, meant that justification made mathematical ideas explicit for all students:

> The practice of justification made space for mathematical discussions that might not otherwise be afforded. Particularly given the broad range of students' prior knowledge, receiving a justification that satisfied an individual was important as explanations were adapted to the needs of individuals, and mathematics that might not otherwise be addressed was brought to the surface. (p. 631)

Also significant is the absence of "ethnic cliques" among the school's population of predominantly poor and marginalized groups, and the fact that the students included mathematics "as part of their futures" (p. 610). Within the context of my central concerns, they had gained access to both a mathematics literacy and positive mathematics identities:

> When we interviewed the students and asked them "what does it take to be successful in mathematics class?" they offered many different practices such as: asking good questions, rephrasing problems, explaining well, being logical, justifying work, considering answers, and using manipulatives. . . . Railside students regarded mathematical success much more broadly than students in the traditional classes, and instead of viewing mathematics as a set of methods that they needed to observe and remember, they regarded mathematics as a way of working with many different dimensions. . . . Put simply, *when there are many ways to be successful, many more students are successful.* (pp. 629–30)

Developing Identities of Inclusion

The research reviewed in this chapter underlines the complexity of the issues involved in understanding how mathematics identities develop and what is involved in addressing the issues of exclusion and non-participation. In focusing on the concept of mathematical literacy, I have aimed to keep the nature of mathematical discourse clearly in view, and to forefront the twin issues of differential access to the discourse and its role in sense-making. Mathematical sense-making itself has a number of dimensions: neatly captured in Brenner's framework, it ranges from the generation of new meanings in the use of technical vocabulary and metaphor, through the discursive practices of its knowledge claims and their justification, to its use in application in order to make sense of other phenomena. In this latter sense, access to mathematical literacy is clearly a social justice issue, but inequity exists on many levels: class and culture interact with the sense-making structures of mathematics to result in differential access to it in terms of both linguistic and cultural capital. Combined with deeply embedded discourses of ability which include assumptions about gender and ethnicity, the subtleties of such differential access serve to generate particular subject positions and personal epistemologies which afford only a restricted view of mathematics to many students. The development of an identity of inclusion, in terms of a view of oneself as legitimate peripheral participant, requires both cultural resources and a "history in person" which enables either take-up of favorably offered positions, or resistance and refiguring when ascribed identities are excluding.

Hence I have also argued that access to a mathematics literacy requires its invisible practices to be made visible. This presents a particular challenge, because in acknowledging the power of mathematical discourse, we also imbue the teacher-expert with considerable power and authority. This presents a pedagogical problem because, as I argued in Chapter 8, doing mathematics involves knowing *how* rather than knowing *that* and so necessarily involves participating in the established practice of mathematics, rather than re-inventing it or negotiating it for oneself—Sfard grapples with this problem in terms of the interpretation of NCTM *Standards*, as does Ball in her dilemmas of teaching elementary school mathematics—how do we respect students' thinking while maintaining intellectual honesty? As Sfard acknowledges, there is no easy answer to this, and I have invoked the concepts of epistemic versus social authority, and mutual trust in my own particular response. These considerations are reflected in the research and projects outlined in this chapter—in enabling access to mathematical literacy, they are all dependent on a re-figuring of classroom relationships to include an explicit focus on investigation, explanation and justification in mathematics, not only as key related processes in the practice of mathematics, but also as essential components of an equitable, relational pedagogy.

Writing from a British perspective, and drawing on data from students in England, the issue of an inclusive mathematics literacy seems all the more pressing within its climate of audit, assessment and ability discourses. Perhaps the biggest challenge to these discourses comes from the fact that even successful students feel disempowered with respect to mathematics. This is best articulated by Sue, a first-year undergraduate from Bradley who has chosen to study mathematics at university, and who is not alone in her feeling that mathematics is a mystery outside of her own sphere of understanding—or making—when she says that "In maths it seems they change the rules when they want . . . Why don't they just tell you the truth?"

Appendix A
The Education System in England and Wales

This book draws on data gathered in English schools and universities, and some explanation will be needed for readers who are unfamiliar with this system. The most important feature of it for this book is the background of audit and assessment in England and Wales. I also provide some notes on the structure of the system in terms of student movement through it.

The Education Structure

Education is compulsory between the ages of 5 and 16, and children start school at the latest in the term in which they turn 5, but often in the school year in which they turn 5—in practice this means that many children, starting in September, may be only a little over 4 years old. The first year of school is called Reception, followed by years 1 to 11 (age 15–16). Although some areas of the country operate a middle school system—primary (age 4–8), middle (age 8–13) and upper (age 13–16/18)—during these compulsory years, most areas have a system of primary school from Reception to Year 6 (age 10–11), followed by secondary school from Year 7 to Year 11. The schools in this book followed this pattern. At the end of Year 11, young people can choose to stay on in full-time education for a further two years, and the majority do in fact do so. These are Years 12 and 13, although they are often called "sixth form," a hangover from an earlier era. Where these two further years take place varies: many secondary schools have "sixth forms," but there are also sixth form colleges, and also Colleges of Further Education, which offer a wider range of programs than schools, including vocational and part-time programs.

Although many areas of England and Wales introduced "comprehensive" (i.e. non-selective) secondary school systems during the 1960s and 1970s, selective schools were preserved in some areas, and so these "grammar schools" exist alongside comprehensive schools in some towns. Originally part of the post-war education system which also included "technical" schools and "secondary modern" schools, grammar schools are designed to teach high academic attainers, selecting their pupils by means of the "11-plus" examination, taken during Year 6. Depending on their location, the effect of grammar schools can be to "cream off" children from economically advantaged backgrounds, leaving competing comprehensives with a rather different population than they might have had. The primary school described in Chapter 2 is located in a town with this kind of grammar school system. Northdown School, on the other hand, is a comprehensive school in a comprehensive area.

Following Years 12 and 13, a large number of young people will move on to university, where they will typically study for three years (four in Scotland) to take a BA or BSc degree. Degree classification is in terms of first class honors, upper second class honors, lower second class honors, third class honors, and Pass, and most students will aim for an upper second (marks falling in the 60–69% range). British universities fall into clear groupings which students are generally aware of: these are the old established elite universities; the "red brick" universities founded in the nineteenth and early twentieth centuries; the "plateglass" universities, founded as a result of the expansion of university education in the 1960s; and finally the "new" universities, which are polytechnics and colleges granted university status after 1992. Bradley and Middleton are "plateglass" universities, while Farnden is a "new" university. New universities differ from the others in that they tend to receive far less funding for research: as a result, teaching loads are often higher in these institutions. They tend to be less prestigious and to attract students with lower entry qualifications, often coming from local communities. University staff are interchangeably called lecturers or tutors by the students.

Assessment and Audit

In English and Welsh schools, the school curriculum is divided up into "Key Stages": Key Stage 1 (Years 1–2), Key Stage 2 (Years 3–6), Key Stage 3 (Years 7–9), Key Stage 4 (Years 10–11), and Key Stage 5 (Years 12–13). Standard Assessment Tests (SATs) are established as part of an audit system which publicizes "league tables" of schools in terms of pupil achievement in literacy, numeracy, and science at the ends of Key Stages 1, 2, and 3 (Years 2, 6, and 9). The tests are designed so that most pupils will progress by approximately one level every two years, and the expectation is that the majority of pupils will gain a Level 2 at the age of 7 and Level 4 at the age of 11 (77% gained Level 4 or above in mathematics in 2007, 33% achieving Level 5; DCSF, 2007a). While these were originally developed as a measure of school performance, they have increasingly been treated as measures of individual performance, with reported detrimental effects, especially for younger children (see for example Reay & Wiliam, 1999). Although statutory testing takes place only in Years 2, 6, and 9, most schools carry out non-statutory tests in other years. They also carry out a range of other tests; the primary school in this study used the Richmond Tests of study skills, language, reading, and mathematics (Hieronymus, Lindquist, & France, 1988) in Year 5. This test had local currency with respect to acceptance at certain secondary schools in Year 7. Alongside this focus on assessment, there have been two major interventions in the primary school curriculum and the way it is delivered. The first of these was the National Literacy Strategy, launched in 1997; it established a system of daily lessons on literacy—"the literacy hour"—with highly structured curriculum aims and recommended whole-class teaching for half of the lesson, followed by small group work sessions. Modeled on the Literacy Strategy, the National Numeracy Strategy was established in 1999, and is similarly known by teachers, parents, and pupils alike as the "numeracy hour"; it is driven in both form and content by the remit of "raising standards" through a daily hour of mathematics teaching which again is highly structured with respect to both curriculum and teaching style (DCSF, 2007b). The National Strategy was introduced in secondary schools at Key Stage 3 in 2001.

At the end of Key Stage 4, in Year 11, students take the General Certificate of Secondary Education Examinations (GCSEs). These are graded from A to G, with an A* grade for exceptional performance and a U grade for a fail. Although all grades apart from U are technically passes, the major "currency" in GSCE examinations is in terms of how many passes one has at grades A* to C. Schools are ranked in publicly available league tables according to how many students achieve grades in this range. A major usage of these league tables is in the process of secondary school choice, whereby parents choose schools for their children on the basis of school performance. Hence GCSE performance becomes a major issue for schools and teachers. Students typically take 10 GCSE subjects, and mathematics, English, and science are compulsory. Many GCSE examinations are "tiered" in that students enter for a certain level and curriculum coverage, meaning that those entered for the Higher Tier can achieve up to an A* grade, while those entered for the Foundation Tier can at best receive a C grade. Mathematics GCSE until very recently was assessed on a 3-tier system, and the change has been made not without controversy. In sixth form, students can choose both academic and vocational courses. Academic courses are more prestigious, and are the major currency for entry to degree-level study. Students have a considerable amount of choice in what they study, typically taking four subjects at AS level in Year 12, and continuing with three of these into Year 13, when they take their final A level examinations.

Appendix B
The Students, the Data and the Analysis

The data on which this book is based are drawn from a number of studies and I describe the samples, the nature of the data and my analysis here. I have used pseudonyms for all people and places.

The Primary School Children

The full data set includes recordings of 24 hour-long mathematics lessons spread over a period of 5 months, and interviews with pupils and their teacher. I draw in this book on Black's original categorization of the children into groups as shown in Table 1.

The Secondary School Students

Participants were selected for interview by the head of the mathematics department, who was asked to pick pupils at random from the roll. There is a bias towards higher ability group students in the sample, a fact which may reflect the Head of Mathematics' anxiety about the schools' performance in her selection of students. However, two of the higher group students had in fact been moved up from lower groups at the end of the previous year and draw explicit contrasts between them. Additionally, this spread allowed me to explore complexities within the higher ability group students' developing identities and to provide a stronger contrastive basis for analysis of the undergraduate group data.

Pupils were interviewed on a one-to-one basis for approximately half an hour each. The format of the interview was open-ended, and covered the following topics: likes and dislikes in mathematics and other school subjects; the experience of SATs at Key Stage 2 (end of Year 6, before entry into their current school) and Key Stage 3 (end of Year 9); ability grouping; effort and ability; creativity in mathematics; the experience of learning mathematics; working in groups; using mathematics in the real world; and differences between boys and girls in mathematics classrooms.

Table 1: The primary school children grouped in terms of productive and non-productive interactions, adapted from (Black, 2004a, pp. 37–9)

Pupil	Productive interactions[2]	Non-productive interactions	Total no. of interactions
Group A			
Toby	10	5	15
Jeremy[1]	11	5	16
Phillip	30	22	52
Simon	18	12	32
Sean	40	22	64
Tim	19	7	31
Chris	7	3	12
Daniel	32	24	56
Group B			
Nelson	8	13	22
Jason	7	17	28
Hasan	4	6	17
Carl	2	4	11
Erica	10	13	23
Group C			
Jennifer	3	4	10
Jane	2	2	6
Alice	3	3	7
Rosie	3	2	5
Chantel	1	2	3
Rachel	3	6	9
Angela	0	2	3
Samantha	0	2	4
Paul	0	0	2
Group D			
Joanne	5	4	12
Kathryn	7	6	14
James	21	20	41
Sian	9	10	21
Peter	12	12	28
Ben	8	9	18

[1] Left half way through the period of observation.
[2] For details of the analytic framework used by Black to place children into talk-type categories, see Black (2004c).

Table 2: The secondary school students grouped by year and ability group membership

Student name	Male/Female	Year of study	Ability group membership
Carol	F	7	Within-class lower
Beccy	F	7	Within-class lower
Sylvia	F	7	Within-class lower
David	M	7	Within-class lower
John	M	7	Within-class lower
Nicola	F	7	Within-class higher
Jonathan	M	7	Within-class higher
Jake	M	7	Within-class higher
Trevor	M	9	Lower
Lizzie[1]	F	9	Lower
Luke	M	9	Higher
Mark	M	9	Higher
Michael	M	9	Higher
Daniel	M	9	Higher
Jenny	F	9	Higher
Georgia	F	9	Higher
Rosie	F	9	Higher
Ben	M	10	Lower
Paul	M	10	Lower
Anna	F	10	Lower
Tom	M	10	Lower
Rachel[2]	F	10	Higher
Sue	F	10	Higher (GCSE yr 10)
Harry	M	10	Higher (GCSE yr 10)
Kate[2]	F	10	Higher
Christopher	M	10	Higher

[1] Moved down from higher group at beginning of the year.
[2] Moved up from lower group at beginning of the year.

The Undergraduates

As outlined in Chapter 1, these were drawn from three different universities.

The First Years

The first-year students were drawn from Bradley University, and were self-selecting, having responded to a request delivered via their tutors to help with a project concerning mathematics learning in which they would get an opportunity to talk about their own study experiences. Ten respondents were aged 19–20, and included four women and six men; the eleventh was a 23-year-old male mature student, and the twelfth was a 34-year-old female mature student. Schools in England offer two mathematics qualifications between the ages of 16 and 18: in addition to the standard Advanced Level General Certificate of Education in Mathematics, some students take Advanced Level Further Mathematics, which builds on the material of the standard syllabus. Of the regular age students all had taken Advanced Level Mathematics and one had taken Further Mathematics; both mature students had entered the university with a further education college access award in mathematics. All were taking the basic first-year mathematics course offered at this university, but six were taking an additional mathematics course, compulsory for intending mathematics single majors. Three students (one male, two female) were registered for a single major degree in mathematics, one (female) for a single major in applied mathematics, two (both male) for a joint degree in mathematics combined with computer science, one (male) for a joint degree in mathematics and management, one (female) for a combined sciences degree with mathematics options, and four (one female, three male) for major degrees in other subjects with mathematics as a minor subject.

While all students enter the university in order to study a particular major or joint major degree, a small number opt to change their intended major at the end of their first year, pursuing instead another degree program. Some of the students in the sample were intending to make these sorts of changes. The three mathematics single major students were intending to continue as mathematics majors into the second year of university and the applied mathematics student was moving to environmental science and taking statistics as a minor only. Of the three students who were combining mathematics with another subject as joint majors, one was continuing as joint, one was intending to take mathematics as a minor subject only, and one was intending to change from a joint degree in mathematics and computer science to a single major in mathematics. The remaining students—taking mathematics as a minor or as part of a general science degree—showed a similar variety of intentions. One notable instance was Richard's complete change of major from management to mathematics. Ten of the twelve were planning to continue with mathematics in some form in their degree—only Diane and Charlie were not. These details are summarized in Table 3, which shows each participant's registered major on entry to the university, and their intended major for the second and third years of their degree.

The students were contacted by e-mail and asked to come along to the interview with a selection of work, including a topic they had enjoyed and/or found easy, and a topic which they had disliked or found difficult to do. The interviews were semi-structured, lasting for approximately one hour each and focusing on the following issues: the students' "mathematics histories" and comparisons between mathematics at school or college and at university, the effect of different teaching styles on their learning experiences, their experiences of getting "stuck" and strategies for resolving problems, the topics they found easy or hard (students were asked to talk through the examples they had brought with them), comparisons with other subjects in terms of the kind of work expected and how they approached the subject matter and tasks, their reasons for choosing mathematics at university, their views on what kind of approach

Table 3: First-year undergraduate profiles: Bradley

Student name	Male/ Female	Registered major on entry: Mathematics majors/joint majors are in bold	Intended second and third year major subjects: Mathematics majors/joint majors are in bold
Carol	F	**Environmental mathematics**	Mathematics minor only
Debbie (mature)	F	**Single major mathematics** Religious Studies minor	**Single major mathematics**
Sarah	F	**Single major mathematics** Art minor	**Single major mathematics**
Larry	M	**Single major mathematics**	**Single major mathematics**
Pete (mature)	M	**Mathematics/computer science joint**	**Mathematics/computer science joint**
Steve	M	**Mathematics/computer science joint**	**Single major mathematics**
Joe	M	**Management/mathematics joint**	Statistics minor only
Sue	F	Combined sciences (includes mathematics options)	Combined sciences, including mathematics
Diane	F	Geography	Geography
Charlie	M	Computer science	Communication studies
Chris	M	Natural sciences	Statistics minor only
Richard	M	Management	**Single major mathematics**

would lead to success in mathematics, and their perceptions of research mathematics and of themselves as mathematicians. The students were interviewed individually when they were approximately two-thirds of the way through their first year at university. The interviews were audio-taped.

The Second and Third Years

The data from the second and third year students were collected in two closely connected locations, when the students were five weeks into the academic year (Farnden) and seven weeks in (Middleton). The two universities are geographically very close, and there are strong connections between the two in terms of an extensive mathematics support program which is aimed not just at mathematics students but also at those who are studying mathematically-demanding degrees such as engineering or nursing. The students were recruited to the focus groups/joint interviews by their tutors in teaching sessions. All the students were of regular undergraduate age, and included eight women and seven men. All but two were studying single major degrees in mathematics. These details are summarized in Table 4.

The interviews/focus groups were semi-structured, and lasted for approximately one hour. Students were asked to discuss what they enjoyed and did not enjoy about mathematics, what they felt about the transition from the first year to years 2 and 3 work, and how they had responded to change. They were also asked to comment on their experiences of lectures, tutorials and tutor support, how they coped with difficulty, and in particular their views on the mathematics support centers and the kinds of students who used these. They were also asked to discuss the mathematics student culture, specifically its work culture, and their aspirations for the future. The groups were audio-taped.

Analyzing the Data

Observation data: For the fullest account of this data and its analysis see Black (2004c). I will briefly summarize here the processes involved in the generation of the talk-type groupings in Table 1. The observation data were transcribed in full before analysis. A content analysis based on Edwards and Mercer's (1987) and Barnes' (1976) theorizations of classroom talk was used to code exchanges in terms of open/closed questions, cued elicitation, recognition of pupil contributions as valid/invalid and so on. The focus was on the net result of any particular exchange as productive or unproductive, defined as creating and

Table 4: Second- and third-year undergraduate profiles (Farnden and Middleton)

Student name	Male/female	Year of study	Location	Degree scheme
Ros	F	2	Farnden	Mathematics
Caitlin	F	2	Farnden	Mathematics
Tamsin	F	2	Farnden	Mathematics and Statistics
Liz	F	3	Farnden	Mathematics and Accounting
Rachel	F	3	Farnden	Mathematics
Jess	F	2	Middleton	Mathematics
Emma	F	2	Middleton	Mathematics
Megan	F	2	Middleton	Mathematics
Chun	M	2	Middleton	Mathematics
Tim	M	2	Middleton	Mathematics
Matt	M	2	Middleton	Mathematics
James	M	2	Middleton	Mathematics
Nick	M	2	Middleton	Mathematics
Yu	M	2	Middleton	Mathematics
Liang	M	2	Middleton	Mathematics

maintaining shared mental context, or preventing achievement of a shared mental context. A cumulative analysis of the patterns of each child's exchanges as productive or unproductive was used to generate an overall categorization into the groups in Table 1.

Interview and focus group data: My analysis of the interviews began with a literal reading of each transcript in which I observed particular features, for example the use of particular words and language. I used these observations to generate categories which could be used to code the data and make comparisons between individuals or groups of individuals. These early categories included:

- motivation patterns and learning styles
- negative learner identity
- observations on learning
- parental involvement
- passive learner
- pedagogic processes
- performance orientation
- practice awareness
- proof
- purpose of teacher's questions
- talking to other students
- teaching style
- maths is hard
- attitude to maths negative
- attitude to maths positive
- "boring"
- comparison other students
- definitions of maths
- functioning in class
- identity
- instrumental maths use
- lack of ownership
- mastery orientation
- maths comparisons with other subjects.

I was also informed by the theoretical frameworks outlined in Chapter 1, and I coded the data in terms of particular categories that I wanted to explore in a more interpretive reading. Thus, using the communities of practice and epistemologies of mathematics frameworks, I focused on participants' relationships with mathematics from a number of perspectives: attitudes (their own and that of others) towards mathematics and definitions of it; comparisons between the students' current experiences of mathematics and their earlier experiences; beliefs about learning (with respect to students' own perceived learning styles but also with respect to mathematics learning generally); performance versus mastery orientations; memory versus understanding; relationships with teachers; experiences of teaching styles; and negative or positive emotions. Repeated exploration of these categories and the connections between them revealed emergent themes and thus further complexities in the students' positionings of self. These themes indicated issues of importance in their experiences and I used these for further exploration of the data set; in particular I looked for connections between categories within the same interview, and I made comparisons between participants with respect to the same categories. One such emergent theme was the importance of gender and its cross-cutting of identities of engagement and alignment in the first-year undergraduate group, and of ability grouping and engagement in the secondary school students. (See Mason, 2002; Seale, 2000, for an analysis of techniques similar to those employed here.)

Notes

1   Some definitions and terms in Systemic Functional Linguistics are helpful, and I draw on Schleppegrell (2004, pp. 71ff) and also Christie (1998) here.

Nominalization: this is the process by which verbs and adjectives become nouns/noun phrases, for example *tend* → **tendency**; *classify* → **classification**; *abstract* → **abstraction**. The effect of nominalization is to condense text, and it is frequently used in academic texts. Schleppegrell cites Ravelli (1996):

> Nominalisation is usually associated with other, related, linguistic features including complex nominal group structure, with many pre and post modifiers, the use of embedded clauses, and lexical choices which are prestigious, technical and formal, rather than coming from a more everyday realm. (p. 380)

Grammatical metaphor: makes use of nominalization to express concepts in an *incongruent* form, that is, in a way which does not map onto the everyday, *congruent* expression in which " 'things' are realized in nouns, 'happenings' are realized in verbs, 'circumstances' are realized in adverbs or prepositional phrases, and relations between elements are realized in conjunctions" (Schleppegrell, p. 72). What is important about grammatical metaphor is that it does not map elements and grammatical categories in this way. Nominalization assists by presenting actions as nouns, as in "*The telephone was invented*" → "**The invention of the telephone**." As metaphor, it differs from lexical metaphor, which uses the same term to convey different meaning (e.g. "toothless rage"), instead using different grammatical categories to convey the same meaning. What is crucial about grammatical metaphor is that it is a way of *making meaning*:

> Through grammatical metaphor, "everyday" meanings are construed in new ways that enable the abstraction, technicality, and development of arguments that characterize advanced literacy tasks. (Schleppegrell, p. 72)

It does this by enabling chains of reasoning and argument which would be difficult to present congruently. Another way in which this functions is by representing conjunctive relations such as "because" as processes, with the effect of shifting the locus of reasoning to within a single clause rather than between clauses. Thus, to take another example from Schleppegrell (p. 73), we can shift the congruent sentence "*Because the telephone was invented, there were many new opportunities for better communication*" into the incongruent "**The invention of the telephone created many opportunities for enhanced communication**," by replacing "because" with the verb "create." Reasoning is thus enabled by the verb rather than the conjunction, and this is a common feature of academic or specialized registers.

Register: relates to the situation and its related lexical choices. Christie (1998, p. 53) notes that if we compare shopping and writing a report, their different situational contexts require correspondingly different language choices: they denote different interpersonal relationships, different content, and different text use (in this case, face to face versus written remote).

Genre: relates to the cultural context and the language choices made in a "staged, purposeful, goal-directed activity" (Christie, pp. 53–4). Thus there is a "shopping genre," and a "report genre," that is, stages through which participants must move to achieve their goals in that activity. For example, the narrative genre involves moving through the stages of abstract, orientation, complication, evaluation, resolution, coda (see Schleppegrell, p. 83). Personal genres include recount and narrative, factual genres include procedure and report, and analytical genres include account, explanation, and exposition; each has distinctive register features (see Schleppegrell, pp. 84–5).

2   I am grateful to Laura Black for allowing me access to her data and giving me permission to quote from it extensively in this book. I have drawn on her initial analysis of the data set in terms of individual children's membership of talk-type groupings and used this as a basis for my commentary.

3   Hamilton even wrote poems about quaternions, as in "The Tetractys," written to J. Herschel:

> **The Tetractys**
> On high Mathesis, with its "Charm severe
> Of line and number," was our theme, and we
> Sought to behold its unborn progeny,
> And thrones reserved in Truth's celestial sphere,
> While views before attained became more clear:
> And how the One of Time, of Space the Three,
> Might in the Chain of Symbols girdled be.
> And when my eager and revented ear
> Caught some faint echoes of an ancient strain,
> Some shadowy outline of old thoughts sublime,
> Gently He smiled to mark revive again,
> In later age, and occidental clime,
> A dimly traced Pythagorean lore;
> A westward floating, mystic dream of FOUR.
> "OBSERVATORY, October 1846"

(Graves, 1885, p. 525)

The poem is not a presentation *of* mathematics. It is *about* mathematics.

4   I am grateful to Tamara Bibby for giving me permission to use these data which formed the focus of her presentation at this seminar.

References

AAUW. (1995). *How schools shortchange girls.* Washington: American Association of University Women Educational Foundation.

Adler, J. (1997). A participatory-inquiry approach and the mediation of mathematical knowledge in a multilingual classroom. *Educational Studies in Mathematics, 33,* 235–258.

Alexander, P., & Dochy, P. (1995). Conceptions of knowledge and beliefs: A comparison across varying cultural and educational communities. *American Educational Research Journal, 32,* 413–442.

Alibert, D., & Thomas, M. (1991). Research on mathematical proof. In D. Tall (Ed.), *Advanced mathematical thinking.* Dordrecht: Kluwer.

Almeida, D. (2000). A survey of mathematics undergraduates' interaction with proof: Some implications for mathematics education. *International Journal of Mathematical Education in Science and Technology, 31*(6), 869–890.

Anderson, J. (1996). The legacy of school—attempts at justifying and proving among new undergraduates. *Teaching Mathematics and its Applications, 15*(3), 129–134.

Augustine, D. (1977). On Order II, 9.26. In V. Bourke (Ed.), *The essential Augustine.* Indianapolis: Hackett.

Back, J., & Pratt, N. (2007). *Spaces to discuss mathematics: Communities of practice on an online discussion board.* Paper presented at the 2nd Socio-cultural Theory in Educational Research and Practice Conference: Theory, Identity and Learning, Manchester University, September.

Baker, D., Clay, J., & Fox, C. (1996). *Challenging ways of knowing in English, Maths and Science.* London: Falmer.

Ball, D. L. (1993). With an eye on the mathematical horizon: Dilemmas of teaching elementary school mathematics. *The Elementary School Journal, 93,* 373–397.

Ball, S. J. (1981). *Beachside Comprehensive: A case-study of secondary schooling.* Cambridge: Cambridge University Press.

Barnes, D. (1976). *From communication to curriculum.* Harmondsworth: Penguin.

Barnes, D. (2008). Exploratory talk for learning. In N. Mercer & S. Hodgkinson (Eds.), *Exploring talk in school: Inspired by the work of Douglas Barnes.* London: Sage.

Barnes, D., & Sheeran, Y. (1992). Oracy and genre: speech styles in the classroom. In K. Norman (Ed.), *Thinking voices.* London: Hodder and Stoughton.

Barnes, D., & Todd, F. (1977). *Communication and learning in small groups.* London: Routledge and Kegan Paul.

Barrs, M. (1994). Genre theory: What's it all about? In B. Stierer & J. Maybin (Eds.), *Language, literacy and learning in educational practice.* Clevedon: Multilingual Matters.

Bartholomew, H. (1999). *Setting in stone? How ability grouping practices structure and constrain achievement in mathematics.* Paper presented at the Annual Conference of the British Educational Research Association, University of Sussex, Brighton.

Bartholomew, H. (2000). *Negotiating identity in the community of the mathematics classroom.* Paper presented at the British Education Research Association, Cardiff, Wales.

Bartholomew, H., & Rodd, M. (2003). A fiercely held modesty: the experiences of women studying mathematics. *New Zealand Journal of Mathematics, 32*(3), 9–13.

Barton, D., & Hamilton, M. (1998). *Local literacies: Reading and writing in one community.* New York: Routledge.

Barton, D., Hamilton, M., & Ivanic, R. (Eds.). (2000). *Situated literacies: Reading and writing.* London: Routledge

Barton, D., & Padmore, S. (1994). Roles, networks and values in everyday writing. In D. Graddol, J. Maybin & B. Steirer (Eds.), *Researching language and literacy in social context.* Clevedon: Multilingual Matters.

Barwell, R. (2004). What is numeracy? [Communication] *For the Learning of Mathematics, 24*(1), 20–22.

Barwell, R. (2005). Ambiguity in the mathematics classroom. *Language and Education, 19*(2), 118–126.

Bauersfeld, H. (1995). The structuring of structures: Development and function of mathematizing as a social practice. In L. Steffe & J. Gale (Eds.), *Constructivism in education* (pp. 137–158). Hillsdale, NJ: Lawrence Erlbaum Associates.

Baxter, J. (2002). Jokers in the pack: Why boys are more adept than girls at speaking in public settings. *Language and Education, 16*(2), 81–96.

Baxter Magolda, M. (2002). Epistemological reflection: the evolution of epistemological assumptions from age 18 to 30. In B. Hofer & P. R. Pintrich (Eds.), *Personal epistemology: The psychology of beliefs about knowledge and knowing.* London: Lawrence Erlbaum Associates.

Becker, J. R. (1995). Women's ways of knowing in mathematics. In G. Kaiser & P. Rogers (Eds.), *Equity in mathematics education: Influences of feminism and culture* (pp. 163–174). London: Falmer.

Belenky, M., Clinchy, B., Goldberger, N., & Tarule, J. (1986). *Women's ways of knowing: The development of self, voice and mind.* New York: Basic Books.

Bendixen, L. D. (2002). A process model of epistemic belief change. In B. Hofer & P. R. Pintrich (Eds.), *Personal epistemology: The psychology of beliefs about knowledge and knowing.* London: Lawrence Erlbaum Associates.

Bernstein, B. (1990). *The structuring of pedagogic discourse, Volume IV: Class, codes and control.* London: Routledge.

Bernstein, B. (1996). *Pedagogy, symbolic control, and identity.* London: Taylor & Francis.

Bernstein, B. B. (1971–75). *Class, codes and control.* London: Routledge and Kegan Paul.

Berry, M. (1981). Systemic linguistics and discourse analysis: A multi-layered approach to exchange structure. In M. Coulthard & M. Montgomery (Eds.), *Studies in discourse analysis* (pp. 120–145). London; Boston: Routledge & Kegan Paul.

Bibby, T. (2002). Shame: An emotional response to doing mathematics as an adult and a teacher. *British Educational Research Journal, 28*(5), 705–721.

Bibby, T. (2006, November). *Mathematical relationships: identities and participation—A psychoanalytic perspective?* Paper presented at the ESRC seminar: Mathematical relationships, Manchester University.

Bibby, T. (forthcoming). In L. Black, H. Mendick & Y. Solomon (Eds.) *Mathematical relationships in education: Identities and participation.* New York: Routledge

Biggs, A., & Edwards, V. (1991). I treat them all the same: Teacher–pupil talk in multiethnic classrooms. *Language and Education, 5*(1), 161–176.

Black, L. (2002a). *The guided construction of educational inequality: How socially disadvantaged children are marginalized in classroom interactions.* Unpublished PhD thesis, Lancaster University, Lancaster.

Black, L. (2002b). *Classroom interaction.* Unpublished raw data.

Black, L. (2004a). Differential participation in whole-class discussions and the construction of marginalised identities. *Journal of Educational Enquiry, 5*(1), 34–54.

Black, L. (2004b, September). *Girls' and boys' participation in whole class discussions and the construction of learner identities in classroom learning processes.* Paper presented at the Annual Conference of the British Educational Research Association, Manchester.

Black, L. (2004c). Teacher–pupil talk in whole-class discussions and processes of social positioning within the primary school classroom. *Language and Education, 18*(5), 347–360.

Black, L. (2005, September). "*She's not in my head or in my body": pupil participation in teacher–pupil interactions in the primary school classroom.* Paper presented at the Conference on Sociocultural Theory in Educational Research and Practice, Manchester University, Manchester, UK.

Black, L., Mendick, H. & Solomon, Y. (Eds.) (forthcoming) *Mathematical relationships in education: Identities and participation*. New York: Routledge

Boaler, J. (1997a). *Experiencing school mathematics: Teaching styles, sex and setting*. Buckingham: Open University Press.

Boaler, J. (1997b). When even the winners are losers: Evaluating the experiences of "top set" students. *Journal of Curriculum Studies, 29*(2), 65–182.

Boaler, J. (1999). Participation, knowledge and beliefs: A community perspective on mathematics learning. *Educational Studies in Mathematics, 40*(3), 259–281.

Boaler, J. (2000). Mathematics from another world: Traditional communities and the alienation of learners. *Journal of Mathematical Behavior, 18*(4), 379–397.

Boaler, J. (2002). *Experiencing school mathematics: Traditional and reform approaches to teaching*. New Jersey: Lawrence Erlbaum Associates.

Boaler, J. (2006). Urban success. A multidimensional mathematics approach with equitable outcomes. *Phi Delta Kappan* (January), 364–369.

Boaler, J. (2007). Paying the price for sugar and spice. In N. S. Nasir & P. Cobb (Eds.), *Improving access to mathematics*. New York: Teachers College.

Boaler, J. (2008). Promoting "relational equity" and high mathematics achievement through an innovative mixed-ability approach. *British Educational Research Journal, 34*(2) 167–194.

Boaler, J., & Greeno, J. G. (2000). Identity, agency, and knowing in mathematics worlds. In J. Boaler (Ed.), *Multiple perspectives on mathematics teaching and learning* (pp. 171–200). Westport, CT: Ablex.

Boaler, J., & Staples, M. (2008). Creating mathematical futures through an equitable teaching approach: The case of Railside School. *Teachers' College Record, 110*(3), 608–645.

Boaler, J., & Wiliam, D. (2001). "We've still got to learn!" Students' perspectives on ability grouping and mathematics achievement. In P. Gates (Ed.), *Issues in mathematics teaching*. London: RoutledgeFalmer.

Boaler, J., Wiliam, D., & Brown, M. (2000). Students' experiences of ability grouping-disaffection, polarisation and the construction of failure. *British Educational Research Journal, 26*(5), 631–648.

Bourdieu, P., & Wacquant, L. J. D. (1992). *An invitation to reflexive sociology*. Cambridge: Polity.

Bousted, M. (1989). Who talks? *English in Education, 23*(3), 41–51.

Breen, C. (2000). *Becoming more aware: Psychoanalytic insights concerning fear and relationship in the mathematics classroom*. Paper presented at the 24th Conference of the International Group for the Psychology of Mathematics Education, Hiroshima, Japan.

Brenner, M. E. (1994). A communication framework for mathematics: Exemplary instruction for culturally and linguistically diverse students. In B. McLeod (Ed.), *Language and learning: Educating linguistically diverse students*. (pp. 233–267). New York: SUNY Press.

Brown, M., & Macrae, S. (2005). *Students' experiences of undergraduate mathematics: Final report*. Retrieved November 24, 2006 from http://www.esrcsocietytoday.ac.uk/ESRCInfoCentre/.

Brown, M., & Rodd, M. (2004). *Successful undergraduate mathematicians: A study of students in two universities*. Paper presented at the 28th Conference of the International Group for the Psychology of Mathematics Education, Bergen, Norway.

Burton, L. (1995). Moving towards a feminist epistemology of mathematics. In G. Kaiser & P. Rogers (Eds.), *Equity in mathematics education: Influences of feminism and culture* (pp. 209–225). London: Falmer.

Burton, L. (1996). Mathematics, and its learning, as narrative—a literacy for the twenty-first century. In D. Baker, J. Clay & C. Fox (Eds.), *Challenging ways of knowing in English, Maths and Science*. Sussex: Falmer.

Burton, L. (1999a). The implications of a narrative approach to the learning of mathematics. In L. Burton (Ed.), *Learning mathematics: From hierarchies to networks* (pp. 21–35). London: Falmer.

Burton, L. (1999b). *Learning mathematics: From hierarchies to networks*. London: Falmer.

Buxton, L. (1981). *Do you panic about maths? Coping with maths anxiety*. London: Heinemann.

Carraher, D., & Schliemann, A. (2002). Is everyday mathematics truly relevant to mathematics education? In M. E. Brenner & J. N. Moschkovich (Eds.), *Everyday and academic mathematics in the classroom* (pp. 131–153). Reston, VA: National Council of Teachers of Mathematics.

Cazden, C. B. (1988/2001). *Classroom discourse: The language of teaching and learning.* Portsmouth, NH: Heinemann.

Chapman, A. (1995). Intertextuality in school mathematics: The case of functions. *Linguistics and Education, 7*(3), 243–362.

Chouliaraki, L. (1998). Regulation in "progressivist" pedagogic discourse: Individualized teacher–pupil talk. *Discourse & Society, 9*(1), 5–32.

Christie, F. (1998). Learning the literacies of primary and secondary schooling. In F. Christie & R. Misson (Eds.), *Literacy and schooling* (pp. 47–73). London: Routledge.

Christie, F. (2003). *Classroom discourse analysis: A functional perspective.* London: Continuum.

Clewell, B. C., & Campbell, P. B. (2002). Taking stock: Where we've been, where we are, where we're going. *Journal of Women and Minorities in Science and Engineering, 8,* 255–284.

Coates, J. (1997). One-at-a-time: The organization of men's talk. In S. Johnson & U. H. Meinhoff (Eds.), *Language and masculinity.* Oxford: Blackwell.

Cobb, P., & Hodge, L. (2002). A relational perspective on issues of cultural diversity and equity as they play out in the mathematics classroom. *Mathematical Thinking and Learning, 4*(2&3), 249–284.

Cobb, P., & Yackel, E. (1993, October 11–15). *A constructivist perspective on the culture of the mathematics classroom.* Paper presented at The Culture of the Mathematics Classroom, Osnabruck, Germany.

Cobb, P., & Yackel, E. (1998). A constructivist perspective on the culture of the mathematics classroom. In F. Seeger, J. Voigt & U. Waschescio (Eds.), *The culture of the mathematics classroom* (pp. 158–190). Cambridge: Cambridge University Press.

Cobb, P., Yackel, E., & Wood, T. (1993). Discourse, mathematical thinking, and classroom practice. In E. Forman, N. Minick & C. Stone (Eds.), *Contexts for learning: Sociocultural dynamics in children's development* (pp. 91–119). Oxford, England: Oxford University Press.

Cooper, B. (2001). Social class and "real-life" mathematics assessments. In P. Gates (Ed.), *Issues in mathematics teaching.* London: Routledge.

Cooper, B., & Dunne, M. (2000). *Assessing children's mathematical knowledge.* Buckingham: Open University Press.

Corson, D. (1992). Language, gender and education: A critical review linking social justice and power. *Gender & Education 4*(3), 229–255.

Cox, W. (2001). On the expectations of the mathematical knowledge of first-year undergraduates. *International Journal of Mathematical Education in Science and Technology, 32*(6), 847–861.

Crawford, K., Gordon, S., Nicholas, J., & Prosser, M. (1994). Conceptions of mathematics and how it is learned: The perspectives of students entering university. *Learning and Instruction, 4,* 331–345.

Creese, A., Leonard, D., Daniels, H., & Hey, V. (2004). Pedagogic discourses, learning and gender identification. *Language and Education, 18*(3), 191–206.

Czerniewska, P. (1992). *Learning about writing.* Oxford: Blackwell.

Dalton, B., Ingels, S. J., Downing, J., & Bozick, R. (2007). *Advanced mathematics and science coursetaking in the Spring High School senior classes of 1982, 1992, and 2004 (NCES 2007–312).* Retrieved February 11, 2008 from http://nces.ed.gov/pubs2007/2007312.pdf.

Dart, B., & Clarke, J. (1988). Sexism in schools: A new look. *Educational Review, 40,* 41–49.

Davies, B. (1990). Agency as a form of discursive practice: A classroom scene observed. *British Journal of Sociology of Education, 11*(3), 341–361.

Davis, F., West, M. M., Greeno, J. G., Gresalfi, M., & Martin, H. T. (2007). Transactions of mathematical knowledge in the Algebra Project. In N. S. Nasir & P. Cobb (Eds.), *Improving access to mathematics* (pp. 69–88). New York: Teachers College Press.

DCSF. (2007a). *National Curriculum assessments at Key Stage 2 in England, 2007 (Provisional).* Retrieved from http://www.dcsf.gov.uk/rsgateway/DB/SFR/s000737/index.shtml.

DCSF. (2007b). *Primary Framework for literacy and mathematics.* Retrieved from http://www. standards.dfes.gov.uk/primaryframeworks/.

de Abreu, G., Bishop, A., & Pompeu, G. (1997). What children and teachers count as mathematics. In T. Nunes & P. Bryant (Eds.), *Learning and teaching mathematics: An international perspective* (pp. 233–264). Hove: Psychology Press.

de Abreu, G., & Cline, T. (2003). Schooled mathematics and cultural knowledge. *Pedagogy, Culture and Society, 11*(1), 11–30.

de Abreu, G., & Cline, T. (2007). Social valorization of mathematical practices: The implications for learners in multicultural schools. In N. S. Nasir & P. Cobb (Eds.), *Improving access to mathematics.* New York: Teachers College Press.

De Corte, E., Op't Eynde, P., & Verschaffel, L. (2002). "Knowing what to believe": the relevance of students' mathematical beliefs for mathematics education. In B. Hofer & P. Pintrich (Eds.), *Personal epistemology: The psychology of beliefs about knowledge and knowing.* Mahwah, NJ: Lawrence Erlbaum.

Delamont, S. (1990). *Sex roles and the school.* London: Methuen.

DfES. (2000). *The National Numeracy Strategy: Mathematical vocabulary.* London: Department for Education and Skills.

DfES. (2005). Standards Unit Newsletter [Electronic Version], Autumn from http://www.dfes. gov.uk/successforall/downloads/sunewsletterautumn2005–194–318.pdf.

Dillon, J. T. (1982). The multidisciplinary study of questioning. *Journal of Education Psychology, 74,* 147–165.

Dowling, P. (1998). *The sociology of maths education: Mathematical myths/pedagogic texts.* London: Falmer.

Dowling, P. (2001). Reading mathematics texts. In P. Gates (Ed.), *Issues in mathematics teaching.* London: Routledge.

Dreyfus, T. (1999). Why Johnny can't prove. *Educational Studies in Mathematics, 38*(1–3), 23–44.

Durkin, K., & Shire, B. (1991). Lexical ambiguity in mathematical contexts. In K. Durkin & B. Shire (Eds.), *Language in mathematical education* (pp. 71–84). Buckingham: Open University Press.

Dweck, C. (1986). Motivational processes affecting learning. *American Psychologist, 41*(10), 1040–1048.

Eccles, J. (1989). Bringing young women to math and science. In M. Crawford & M. Gentry (Eds.), *Gender and thought: Psychological perspectives* (pp. 36–58). New York: Springer-Verlag.

Edwards, D., & Mercer, N. (1987). *Common knowledge.* London: Routledge.

Ernest, P. (1991). *The philosophy of mathematics education.* Basingstoke: Falmer.

Ernest, P. (1998). The relation between personal and public knowledge from an epistemological perspective. In F. Seeger & J. W. Voigt, (Eds.), *The culture of the mathematics classroom* (pp. 245–268). Cambridge: Cambridge University Press.

Ernest, P. (1999). Forms of knowledge in mathematics and mathematics education: Philosophical and rhetorical perspectives. *Educational Studies in Mathematics, 38*(1–3), 67–83.

Esterhuysen, P. (1996). "Focusing on the frames": Using comic books to challenge dominant literacies in South Africa. In D. Baker, J. Clay & C. Fox (Eds.), *Challenging ways of knowing in English, Maths and Science.* Sussex: Falmer.

Evans, J. (2000). *Adults' mathematical thinking and emotions: A study of numerate practices.* London: RoutledgeFalmer.

Fennema, E., & Leder, C. (Eds.). (1990). *Mathematics and gender: Influences on teachers and students.* New York: Teachers College.

Fennema, E., & Romberg, T. A. (1999). *Mathematics classrooms that promote understanding.* New Jersey: LEA.

Fischbein, E., & Kedem, I. (1982). *Proof and certitude in the development of mathematical thinking.* Paper presented at the Sixth Annual Conference of the International Group for the Psychology of Mathematics Education, Antwerp.

Forman, E. A., Larreamendy-Joerns, J., Stein, M., & Brown, C. A. (1998). "You're going to want to find out which and prove it": Collective argumentation in a mathematics classroom. *Learning and Instruction, 8*(6), 527–548.

Fortescue, C. M. (1994). Using oral and written language to increase understanding of math concepts. *Language Arts, 71*(8), 576–580.

Franke, M., Kazemi, E. & Battey, D. (2007) Mathematics teaching and classroom practice. In F. Lester (Ed.) *Second handbook of research on mathematics teaching and learning* (pp. 225–256). Charlotte, NC: Information Age Publishing.

Frankenstein, M. (1995). Equity in mathematics education: Class in the world outside the class. In W. G. Secada, E. Fennema & L. B. Adajian (Eds.), *New directions for equity in mathematics education* (pp. 165–190). Cambridge: Cambridge University Press.

French, J., & French, P. (1984). Gender imbalances in the primary classroom: An interactional account. *Educational Research, 26*(2), 127–136.

Gallego, M., & Hollingsworth, S. (Eds.). (2000). *What counts as literacy.* New York: Teachers College Press.

Gee, J. (1996). *Social linguistics and literacies: Ideology in discourses.* New York: Routledge Falmer.

Gee, J. (1999). *An introduction to discourse analysis: Theory and method.* New York: Routledge.

Gee, J. (2001). Identity as an analytic lens for research in education. *Review of Research in Education, 25,* 99–125.

Gilbert, P. (1994). Authorising disadvantage: Authorship and creativity in the language classroom. In B. Steirer & J. Maybin (Eds.), *Language, literacy and learning in educational practice.* Clevedon: Multilingual Matters.

Gillborn, D., & Mirza, H. S. (2000). *Educational inequality: Mapping race, class and gender.* London: Ofsted.

Gillborn, D., & Youdell, D. (2000). *Rationing education: Policy, practice, reform, and equity.* Buckingham: Open University Press.

Gilligan, C. (1993). *In a different voice: Psychological theory and women's development.* Cambridge, MA: Harvard University Press.

Good, T., Sikes, N., & Brophy, J. (1973). Effects of teacher sex, student sex on classroom interaction. *Journal of Educational Psychology, 65,* 74–87.

Government Statistical Service. (2007). *Statistics of education: The characteristics of high attainers.* London: HMSO.

Government Statistical Service. (2008). *GCE/VCE A/AS and equivalent examination results in England, 2006/07 (Revised)* (Vol. 2008). London: HMSO.

Graves, R. (1885). *Life of Sir William Rowan Hamilton (Vol.2).* Dublin: Hodges, Figgis and Co.

Grice, H. P. (1975). Logic and conversation. In P. Cole & J. Morgan (Eds.), *Syntax and semantics 3: Speech acts.* New York: Academic.

Griffiths, M. (1995). *Feminisms and the self: The web of identity.* London: Routledge.

Gutiérrez, K. D., Baquedano-López, P., Alvarez, H. H., & Chiu, M. M. (1999). Building a culture of collaboration through hybrid language practices. *Theory into Practice, 38*(2), 87–93.

Gutiérrez, K. D., Rymes, B., & Larson, J. (1995). Script, counterscript, and underlife in the classroom: James Brown versus "Brown v. Board of Education." *Harvard Educational Review, 65*(3), 445–471.

Gutiérrez, R. (2007). (Re)Defining equity: The importance of a critical perspective. In N. S. Nasir & P. Cobb (Eds.), *Improving access to mathematics* (pp. 3–50). New York: Teacher College.

Gutstein, E. (2006). *Reading and writing the world with mathematics: Toward a pedagogy for social justice.* New York: Routledge.

Gutstein, E. (2007). "So one question leads to another": Using mathematics to develop a pedagogy of questioning. In N. S. Nasir & P. Cobb (Eds.), *Improving access to mathematics* (pp. 51–68). New York: Teachers College Press.

Gutstein, E., Lipman, P., Hernandez, P., & de los Reyes, R. (1997). Culturally relevant mathematics teaching in a Mexican American context. *Journal for Research in Mathematics Education, 28*(6), 709–737.

Hall, N. (1994). The emergence of literacy. In B. Steirer & J. Maybin (Eds.), *Language, literacy and learning in educational practice.* Clevedon: Multilingual Matters.

Halliday, M. A. K. (Ed.). (1978). *Language as social semiotic.* London: Edward Arnold.

Halliday, M. A. K. (1994). *An introduction to functional grammar.* London: Arnold.

Hamilton, W. (1843). Quaternions. In H. Halberstam & R. Ingram (Eds.), *The mathematical papers of Sir William Rowan Hamilton Vol. 3 (1967)* (pp. 103–105). Cambridge: Cambridge University Press.

Hamilton, W. (1844). Letter to Graves on quaternions; or on a new system of imaginaries in algebra. *Philosophical Magazine 25*, 489–495.

Hamilton, W. (1844–50). On quaternions; or on a new system of imaginaries in algebra. *Philosophical Magazine* Vols. 25–36. In H. Halberstam & R. Ingram (Eds.), *The mathematical papers of Sir William Rowan Hamilton Vol.3 (1967)* (pp. 227–297). Cambridge: Cambridge University Press.

Hamlyn, D. (1978). *Experience and the growth of understanding.* London: Routledge & Kegan Paul.

Hammer, D., & Elby, A. (2002). On the form of a personal epistemology. In B. K. Hofer & P. Pintrich (Eds.), *Personal epistemology: The psychology of beliefs about knowledge and knowing.* London: Lawrence Erlbaum Associates.

Hammersley, M. (1990). An evaluation of two studies of gender imbalance in primary classrooms. *British Educational Research Journal, 16*, 125–143.

Hanna, G. (1995). Challenges to the importance of proof. *For the Learning of Mathematics, 15*(3), 42–49.

Hannula, M. (2002). Attitude towards mathematics: Emotions, expectations and values. *Educational Studies in Mathematics, 49*, 25–46.

Hardman, F., Smith, F., & Wall, K. (2003). "Interactive whole class teaching" in the National Literacy Strategy. *Cambridge Journal of Education, 33*(2), 197–215.

Hardy, T. (2007). Participation and performance: Keys to confident learning in mathematics? *Research in Mathematics Education, 9.*(eds. L. Bills, J. Hodgen, H. Povey) British Society for Research into the Learning of Mathematics (BSRLM).

Harel, G., & Sowder, L. (1998). Students' proof schemes: Results from exploratory studies. In A. Schoenfeld, J. Kapput & E. Dubinsky (Eds.), *Research in collegiate mathematics education, Vol 3* (pp. 234–283). Providence, RI: American Mathematical Society.

Harel, G., & Sowder, L. (2007). Toward comprehensive perspectives on the learning and teaching of proof. In F. Lester (Ed.), *Second handbook of research on mathematics teaching and learning* (Vol. 2, pp. 805–842). Charlotte, NC: NCTM/Information Age Publishing.

Hasan, R. (2002). Ways of meaning, ways of learning: Code as an explanatory concept. *British Journal of Sociology of Education, 23*(4), 537–548.

Healy, L., & Hoyles, C. (1998). *Justifying and proving in school mathematics: Summary of the results from a survey of the proof conceptions of students in the UK.* London: Institute of Education.

Healy, L., & Hoyles, C. (2000). A study of proof conceptions in algebra. *Journal for Research in Mathematics Education, 31*(4), 396–428.

Heath, S. B. (1983). *Ways with words: Language, life and work in communities and classrooms.* Cambridge: Cambridge University Press.

Heath, S. B. (1996). Good science or good art? or both? In D. Baker, J. Clay & C. Fox (Eds.), *Challenging ways of knowing in English, Maths and Science.* Sussex: Falmer.

Hieronymus, A. N., Lindquist, E. F., & France, N. (1988). *Richmond test of basic skills* (2nd ed.). London: NFER-Nelson.

Higher Education Statistics Agency. (2008). Subject of study 2005–2006. http://www.hesa.ac.uk/index.php?option=com_datatables&Itemid=121, accessed February 11, 2008.

Hill, H. C., Schilling, S. G., & Ball, D. L. (2004). Developing measures of teachers' mathematics knowledge for teaching. *Elementary School Journal, 105*(1), 12–30.

Hofer, B. K. (2000). Dimensionality and disciplinary differences in personal epistemology. *Contemporary Educational Psychology, 25*(4), 378–405.

Hofer, B. K., & Pintrich, P. (1997a). The development of epistemological theories; beliefs about knowledge and knowing and their relation to learning. *Review of Educational Research, 67*, 88–140.

Hofer, B. K., & Pintrich, P. R. (1997b). *Disciplinary ways of knowing: Epistemological beliefs in science and psychology.* Paper presented at the Annual Meeting of the American Educational Research Association, Chicago.

Hofer, B. K., & Pintrich, P. R. (Eds.). (2002). *Personal epistemology: The psychology of beliefs about knowledge and knowing.* London: Lawrence Erlbaum.

Holland, D., Lachicotte Jr, W., Skinner, D., & Cain, C. (1998). *Identity and agency in cultural worlds.* Cambridge, MA: Harvard University Press.

Horn, I. S. (2007). Fast kids, slow kids, lazy kids: Framing the mismatch problem in mathematics teachers' conversations. *Journal of the Learning Sciences, 16*(1), 37–79.

Howe, C. (1997). *Gender and classroom interaction.* Edinburgh: SCRE.

Hull, R. (1985). *The language gap.* London: Methuen.

IES. (2006). *The nation's report card: Trends in average mathematics scale scores by gender.* Retrieved December 18, 2006 from http://nces.ed.gov/nationsreportcard/ltt/results2004/sub-math-gender.asp.

Inhelder, B., & Piaget, J. (1958). *The growth of logical thinking from childhood to adolescence; an essay on the construction of formal operational structures.* New York: Basic Books.

Ireson, J., & Hallam, S. (1999). Raising standards: Is ability grouping the answer? *Oxford Review of Education, 25*(3), 343–358.

Ireson, J., & Hallam, S. (2001). *Ability grouping in education.* London: Paul Chapman.

Ireson, J., Hallam, S., Hack, S., Clark, H., & Plewis, I. (2002). Ability grouping in English secondary schools: Effects on attainment in English, mathematics and science. *Educational Research and Evaluation, 8*(3), 299–318.

Jay, T. (2003). *The psychology of language.* New Jersey: Prentice Hall.

Jenkins, N., & Cheshire, J. (1990). Gender issues in the GCSE oral English examination: Part I. *Language and Education, 4*, 261–292.

Kassem, D. (2001). Ethnicity and mathematics education. In P. Gates (Ed.), *Issues in mathematics teaching* (pp. 64–76). London: Routledge.

Kawanaka, T., & Stigler, J. (1999). Teachers' use of questions in eighth-grade mathematics classrooms in Germany, Japan and the United States. *Mathematical Thinking and Learning, 1*(4), 255–278.

Kenyon, A., & Kenyon, V. (1996). Evolving shared discourse with teachers to promote literacies for learners in South Africa. In D. Baker, J. Clay & C. Fox (Eds.), *Challenging ways of knowing in English, Maths and Science.* Sussex: Falmer.

Khisty, L. (1995). Making inequality: Issues of language and meaning in mathematics teaching with Hispanic students. In W. G. Secada, E. Fennema & L. Adajian (Eds.), *New directions for equity in mathematics education* (pp. 279–297). Cambridge: Cambridge University Press.

King, P. M., & Kitchener, K. S. (1994). *Developing reflective judgement: Understanding and promoting intellectual growth and critical thinking in adolescents and adults.* San Francisco: Jossey-Bass.

Kitchen, A. (1999). The changing profile of entrants to mathematics at A level and to mathematical subjects in higher education. *British Educational Research Journal, 25*(1), 57–64.

Kloosterman, P. (1996). Students' beliefs about knowing and learning mathematics: Implications for motivation. In M. Carr (Ed.), *Motivation in mathematics* (pp. 131–156). Cresskill, NJ: Hampton Press.

Kloosterman, P., & Coughan, M. C. (1994). Students' beliefs about learning school mathematics. *Elementary School Journal, 94*, 375–388.

Kuhn, D., & Weinstock, M. (2002). What is epistemological thinking and why does it matter? In B. Hofer & P. R. Pintrich (Eds.), *Personal epistemology: The psychology of beliefs about knowledge and knowing.* London: Lawrence Erlbaum Associates.

Kyle, J. (2002). Proof and reasoning. In P. Kahn & J. Kyle (Eds.), *Effective learning and teaching in mathematics and its applications.* London: Kogan Page.

Labov, W. (1969). The logic of non-standard English. In J. Alatis (Ed.), *Georgetown monograph on languages and linguistics,* Vol. 22, pp. 1–44.

Ladson-Billings, G. (1995). Making mathematics meaningful in multicultural contexts. In W. Seccada, E. Fennema & L. Adajian (Eds.), *New directions for equity in mathematics education.* Cambridge: Cambridge University Press.

Ladson-Billings, G. (1997). It doesn't add up: African American students' mathematics achievement. *Journal for Research in Mathematics Education, 28*(6), 697–708.

Lakoff, G., & Johnson, M. (1980). *Metaphors we live by*. Chicago: University of Chicago Press.

Lakoff, G., & Nuñez , R. (2001). *Where mathematics comes from: How the embodied mind brings mathematics into being*. New York: Basic Books.

Lampert, M. (1990). When the problem is not the question and the solution is not the answer: Mathematical knowing and teaching. *American Educational Research Journal, 27*, 29–63.

Lampert, M. (2001) *Teaching problems and the problems of teaching*. New Haven, CT: Yale University Press.

Landau, N. R. (1994). *Love, hate and mathematics*. Unpublished MA dissertation, King's College, University of London, London.

Lankshear, C., & Knobel, M. (1998). New times! old ways? In F. Christie & R. Misson (Eds.), *Literacy and schooling*. London: Routledge.

Lave, J., Murtaugh, M., & de la Rocha, O. (1984). The dialectic of arithmetic in grocery shopping. In B. Rogoff & J. Lave (Eds.), *Everyday cognition: Its development in social context* (pp. 67–94). Cambridge, MA: Harvard University Press.

Lave, J., & Wenger, E. (1991). *Situated learning: Legitimate peripheral participation*. Cambridge: Cambridge University Press.

Leder, G. (1995). Equity inside the mathematics classroom: Fact or artifact? In W. G. Secada, E. Fennema & L. Adajian (Eds.), *New directions for equity in mathematics education* (pp. 209–224). Cambridge: Cambridge University Press.

Lee, J., Grigg, W., & Dion, G. (2007). *The nation's report card: Mathematics 2007*. Retrieved from http://nces.ed.gov/pubsearch/pubsinfo.asp?pubid=2007494.

Lemke, J. (1990). *Talking science: Language, learning, and values*. Norwood, NJ: Ablex/Elsevier.

Lemke, J. (2000). Across the scales of time: Artifacts, activities, and meanings in ecosocial systems. *Mind, Culture and Activity 7*(4), 273–290.

Lemke, J. L. (2003). Mathematics in the middle: Measure, picture, gesture, sign, and word. In M. Anderson, A. Sáenz-Ludlow, S. Zellwegr & V. V. Cifarelli (Eds.), *Educational perspectives on mathematics as semiosis: From thinking to interpreting to knowing* (pp. 215–234). Ottawa: Legas.

Lerman, S. (2001). Cultural, discursive psychology: A sociocultural approach to studying the teaching and learning of mathematics. *Educational Studies in Mathematics, 46*(1/3), 87–113.

Lerman, S., & Zevenbergen, R. (2004). The socio-political context of the mathematics classroom: Using Bernstein's theoretical framework to understand classroom communications. In P. Valero & R. Zevenbergen (Eds.), *Researching the socio-political dimensions of mathematics education: Issues of power in theory and methodology* (pp. 27–42). Dordrecht: Kluwer.

Lonka, K., & Lindblom-Ylanne, S. (1996). Epistemologies, conceptions of learning, and study practices in medicine and psychology. *Higher Education Research and Development, 31*, 5–24.

Lubienski, S. T. (2007). Research, reform, and equity in U.S. mathematics education. In N. S. Nasir & P. Cobb (Eds.), *Improving access to mathematics* (pp. 10–23). New York: Teachers College Press.

Ma, X. (2003). Effects of early acceleration of students in mathematics on attitudes toward mathematics and mathematics anxiety. *Teachers College Record, 105*(3), 438–464.

MacGregor, M. (2002). Using words to explain mathematical ideas. *Australian Journal of Language and Literacy, 25*(1), 78–88.

Macken-Horarik, M. (1998). Exploring the requirements of critical school literacy. In F. Christie & R. Misson (Eds.), *Literacy and schooling* (pp. 74–103). London: Routledge.

MacLane, S. (1994). Responses to "Theoretical mathematics: toward a cultural synthesis of mathematics and theoretical physics" by A. Jaffe and F. Quinn. *Bulletin of the American Mathematical Society 30*, 190–193.

MacLure, M., & French, P. (1981). A comparison of talk at home and at school. In G. Wells (Ed.), *Learning through interaction* Cambridge: Cambridge University Press.

Macrae, S., Brown, M., Bartholomew, H., & Rodd, M. (2003a). An examination of one group of failing single honours students in one university. *MSOR Connections, 3*(3), 17–21.

Macrae, S., Brown, M., Bartholomew, H., & Rodd, M. (2003b). *The tale of the tail: An investigation of failing single honours mathematics students in one university.* Paper presented at the British Society for Research into Learning Mathematics, Oxford.

Macrae, S., Brown, M., & Rodd, M. (2001). *Maths is gorgeous: First year students' views of mathematics.* Paper presented at the British Education Research Association, Leeds, England.

Macrae, S., & Maguire, M. (2002). Getting in and getting on: Choosing the "best". In A. Hayton & A. Pacazuska (Eds.), *Access, participation and higher education: Policy and practice.* London: Kogan Page.

Maher, C. A. (2005). How students structure their investigations and learn mathematics: Insights from a long-term study. *Journal of Mathematical Behavior 24,* 1–14.

Marks, R., & Mousley, J. (1990). Mathematics education and genre: Dare we make the process writing mistake again? *Language and Education 4*(2), 117–135.

Martin, D. (2007). Mathematics learning and participation in the African American context: The co-construction of identity in two intersecting realms of experience. In N. S. Nasir & P. Cobb (Eds.), *Improving access to mathematics.* New York: Teachers College Press.

Martin, J. R. (1992). *English text: System and structure.* Amsterdam: Benjamins.

Martin, J. R., Christie, F., & Rothery, J. (1987). Social processes in education. In I. Reid (Ed.), *The place of genre in learning: Current debates* (pp. 58–82). Geelong, Victoria: Deakin University Press.

Martin, J. R., Christie, F., & Rothery, J. (1994). Social processes in education: A reply to Sawyer and Watson (and others). In B. Stierer & J. Maybin (Eds.), *Language, literacy and learning in educational practice.* Clevedon: Multilingual Matters.

Martino, A. M., & Maher, C. A. (1999). Teacher questioning to promote justification and generalization in mathematics: What research has taught us. *Journal of Mathematical Behavior, 18,* 53–78.

Mason, J. (2002). *Qualitative researching* (2nd ed.). London: Sage.

Mendick, H. (2005a). A beautiful myth? The gendering of being /doing "good at maths"." *Gender and Education, 17*(2), 89–105.

Mendick, H. (2005b). Mathematical stories: Why do more boys than girls choose to study mathematics at AS-level in England? *British Journal of Sociology of Education, 26*(2), 225–241.

Mendick, H. (2005c). Only connect: Troubling oppositions in gender and mathematics. *International Journal of Inclusive Education 9*(2), 161–180.

Mendick, H. (2006). *Masculinities in mathematics.* Berkshire: Open University Press.

Mendick, H., Epstein, D., & Moreau, M.-P. (forthcoming). *End of award report: Mathematical images and identities: Education, entertainment, social justice.* Swindon: Economic and Social Research Council.

Mercer, N. (1994). Neo-Vygotskian theory and education. In B. Stierer & J. Maybin (Eds.), *Language, literacy and learning in educational practice.* Clevedon: Multilingual Matters.

Mercer, N. (1995). *The guided construction of knowledge.* Avon: Multilingual Matters.

Mercer, N. (2000). *Words and minds: How we use language to think.* London: Routledge.

Mishler, E. (2000). *Storylines: Craftartists' narratives of identity.* Cambridge, MA: Harvard University Press.

Moore, R. C. (1994). Making the transition to formal proof. *Educational Studies in Mathematics, 27,* 249–266.

Morgan, C. (1998). *Writing mathematically: The discourse of investigation.* London: Falmer.

Morgan, C. (2005). Words, definitions and concepts in discourses of mathematics teaching and learning. *Language and Education, 19*(2), 103–117.

Morgan, C., Tsatsaroni, A., & Lerman, S. (2002). Mathematics teachers' positions and practices in discourses of assessment. *British Journal of Sociology of Education, 23*(3), 445–461.

Morrison, D., & Collins, A. (1996). Epistemic fluency and constructivist learning environments. In B. Wilson (Ed.), *Constructivist learning environments* (pp. 107–119). Englewood Cliffs, NJ: Educational Technology Publications.

Moschkovich, J. N. (2002). An introduction to examining everyday and academic mathematical practices. In M. E. Brenner & J. N. Moschkovich (Eds.), *Everyday and academic mathematics in the classroom* (pp. 1–11). Reston, VA: National Council of Teachers of Mathematics.

Moschkovich, J. N. (2007a). Bilingual mathematics learners: How views of language, bilingual learners, and mathematical communication affect instruction. In N. S. Nasir & P. Cobb (Eds.), *Improving access to mathematics* (pp. 89–104). New York: Teachers College Press.

Moschkovich, J. N. (2007b). Examining mathematical discourse practices. *For the learning of mathematics, 27*(1), 24–30.

Moses, R., & Cobb, C. (2002). *Radical equations: Civil rights from Mississippi to the Algebra Project.* Boston, MA: Beacon Press.

Myhill, D., & Dunkin, F. (2005). Questioning learning. *Language and Education, 19*(5), 415–427.

Nasir, N. S. (2007). Identity, goals, and learning: The case of basketball mathematics. In N. S. Nasir & P. Cobb (Eds.), *Improving access to mathematics* (pp. 132–145). New York: Teachers College Press.

Nathan, M. J., & Knuth, E. J. (2003). A study of whole classroom mathematical discourse and teacher change. *Cognition and Instruction, 21*(2), 175–207.

NCTM. (1991). *Professional standards for teaching mathematics.* Reston, VA: National Council of Teachers of Mathematics.

NCTM. (2000). *Principles and standards for school mathematics.* Reston, VA: National Council of Teachers of Mathematics.

Nimier, J. (1993). Defence mechanisms against mathematics. *For the Learning of Mathematics, 13*(1), 30–34.

Nolder, R. (1991). Mixing metaphor and mathematics in the secondary classroom. In K. Durkin & B. Shire (Eds.), *Language in mathematical education* (pp. 105–113). Buckingham: Open University Press.

Norman, K. (1992). *Thinking voices.* London: Hodder & Stoughton.

Noyes, A. (2007). Mathematical marginalisation and meritocracy: Inequity in an English classroom. *The Montana Mathematics Enthusiast, Monograph 1,* 35–48.

Nunes, T., Schliemann, A. D., & Carraher, D. W. (1993). *Street mathematics and school mathematics.* Cambridge: Cambridge University Press.

O'Halloran, K. (2000). Classroom discourse in mathematics: A multisemiotic analysis. *Linguistics and Education, 10*(3), 359–388.

O'Halloran, K. (2003). Educational implications of mathematics as a multisemiotic discourse. In A. Anderson, A. Sáenz-Ludlow, S. Zellweger & V. V. Cifarelli (Eds.), *Educational perspectives on mathematics as semiosis: From thinking to interpreting to knowing* (pp. 185–214). Legas.

O'Halloran, K. (2005). *Mathematical discourse.* London: Continuum.

O'Halloran, K. (2007). Mathematical and scientific forms of knowledge: A systemic functional multimodal grammatical approach. In F. Christie & J. Martin (Eds.), *Language, knowledge and pedagogy: Functional linguistic and sociological perspectives* (pp. 205–236). London and New York: Continuum.

O'Neill, J. (1993). Intertextual reference in nineteenth-century mathematics. *Science in Context, 6,* 435–468.

Oakes, J. (1990). *Multiplying inequalities: The effects of race, social class, and tracking on opportunities to learn mathematics and science.* Santa Monica, CA: RAND.

Paechter, C. (2001). Gender, reason and emotion in secondary mathematics classrooms. In P. Gates (Ed.), *Issues in mathematics teaching.* London: RoutledgeFalmer.

Perelman, C. (1963). *The idea of justice and the problem of argument.* London: Routledge.

Perry, W. G. (1970). *Forms of intellectual and ethical development in the college years: A scheme.* New York: Holt, Rinehart and Winston.

Perry, W. G. (1999). *Forms of ethical and intellectual development in the college years.* San Francisco: Jossey-Bass.

Picker, S. H., & Berry, J. S. (2000). Investigating pupils' images of mathematicians. *Educational Studies in Mathematics, 43,* 65–94.

Piggott, J. (2004). *Mathematics enrichment: What is it and who is it for?* Paper presented at the Annual Conference of the British Educational Research Association, Manchester.

Pimm, D. (1987). *Speaking mathematically: Communication in mathematics classrooms*. London: Routledge.

Pintrich, P. R., Marx, R., & Boyle, R. (1993). Beyond "cold" conceptual change: The role of motivational beliefs and classroom contextual factors in the process of conceptual change. *Review of Educational Research, 63*(2), 167–199.

Plowden, B. (1967). *Children and their primary schools*. London: Central Advisory Council for Education.

Polanyi, M., & Prosch, H. (1975). *Meaning*. London: University of Chicago Press.

Povey, H., Burton, L., Angier, C., & Boylan, M. (1999). Learners as authors in the mathematics classroom. In L. Burton (Ed.), *Learning mathematics: From hierarchies to networks*. London: Falmer.

Povey, H., Elliott, S., & Lingard, D. (2001). The study of the history of mathematics and the development of an inclusive mathematics: Connections explored. *Mathematics Education Review 14*(September).

Pratt, N., & Kelly, P. (2007). Mapping mathematical communities: Classrooms, research communities and masterclasses. *For the Learning of Mathematics, 27*(3), 34–39.

QCA. (2006). *About mathematics*. Retrieved May 30, 2006 from http://www.qca.org.uk/7883.html

QCA. (2007a). *Mathematics: Programme of study for Key Stage 3 and attainment targets*. Retrieved from http://curriculum.qca.org.uk/subjects/mathematics/keystage3/index.aspx.

QCA. (2007b). *Mathematics: Programme of study for Key Stage 4*. Retrieved from http://curriculum.qca.org.uk/subjects/mathematics/keystage4/index.aspx.

Raman, M. (2001). *Beliefs about proof in collegiate calculus*. Paper presented at the North American Chapter of the International Group for the Psychology of Mathematics Education, Snowbird, Utah.

Raman, M. (2003). Key ideas: What are they and how can they help us understand how people view proof? *Educational Studies in Mathematics, 52*(3), 319–325.

Raman, M., & Zandieh, M. (submitted). The case of Brandon: The dual nature of key ideas in the classroom. *Educational Studies in Mathematics*.

Rav, Y. (1999). Why do we prove theorems? *Philosophia Mathematica, 7*(3), 5–41.

Ravelli, L. (1996). Making language accessible: Successful text writing for museum visitors. *Linguistics and Education, 8*(4), 367–388.

Reay, D., & Wiliam, D. (1999). "I'll be a nothing": Structure, agency and the construction of identity through assessment. *British Educational Research Journal, 25*(3), 343–354.

Recio, A., & Godino, J. (2001). Institutional and personal meanings of mathematical proof. *Educational Studies in Mathematics 48*, 83–99.

Reyes, L. H. (1984). Affective variables and mathematics education. *The Elementary School Journal, 84*(5), 558–581.

Rodd, M. (2002, 1–6 July). *Hot and abstract: Emotion and learning mathematics*. Paper presented at the 2nd Conference on the Teaching of Mathematics, University of Crete, Greece.

Rodd, M. (2003). Witness as participation: The lecture theatre as a site for awe and wonder. *For the Learning of Mathematics, 23*(1), 15–21.

Rodd, M., & Bartholomew, H. (2006). Invisible and special: Young women's experiences as undergraduate mathematics students. *Gender and Education, 18*(1), 35–50.

Rogers, P. (1990). Thoughts on power and pedagogy. In L. Burton (Ed.), *Gender and mathematics: An international perspective* (pp. 38–46). London: Cassell.

Rogers, P. (1995). Putting theory into practice. In P. Rogers & G. Kaiser (Eds.), *Equity in mathematics education: Influences of feminism and culture* (pp. 175–185). London, UK: Falmer.

Rogoff, B. (1990). *Apprenticeship in thinking*. Oxford: Oxford University Press.

Ross, K. (1998). Doing and proving: The place of algorithms and proof in school mathematics. *American Mathematical Monthly, 3*, 252–255.

Ryle, G. (1949). *The concept of mind*. Harmondsworth: Penguin.

Sadker, M., & Sadker, D. (1985). Is the OK classroom OK? *Phi Delta Kappan*, 361.

Sadker, M., & Sadker, D. (1986). Sexism in the classroom: From grade school to graduate school. *Phi Delta Kappan, 68*, 513.

Schleppegrell, M. (2004). *The language of schooling: A functional linguistics perspective*. Mahwah, NJ: Lawrence Erlbaum Associates.

Schleppegrell, M. (2006). The challenges of academic language in school subjects. Retrieved December 5, 2006 from http://www.soe.umich.edu/events/als/downloads/schleppegrellp.html

Schleppegrell, M. (2007). The linguistic challenges of mathematics teaching and learning: A research review. *Reading and Writing Quarterly, 23*(2), 39–159.

Schoenfeld, A. (1988). When good teaching leads to bad results: The disasters of "well-taught" mathematics courses. *Educational Psychologist 23*, 145–166.

Schoenfeld, A. (1989). Explorations of students' mathematical beliefs and behavior. *Journal for Research in Mathematics Education, 20*, 338–355.

Schoenfeld, A. (1992). Learning to think mathematically: Problem-solving, metacognition and sense making in mathematics. In D. A. Grouws (Ed.), *Handbook of research on mathematics teaching and learning* (pp. 334–370). New York: Macmillan.

Schoenfeld, A. (1994). Reflections on doing and teaching mathematics. In Schoenfeld (Ed.), *Mathematical thinking and problem solving* (pp. 53–70). Hillsdale, NJ: Erlbaum.

Schoenfeld, A., & Kilpatrick, A. (2008). Toward a theory of proficiency in teaching mathematics. In D. Tirosh (Ed.), *International handbook of mathematics teacher education: Vol. 2. Tools and processes in mathematics teacher education*. Rotterdam: Sense.

Schommer-Aikins, M. (2002). An evolving theoretical framework for an epistemological belief system. In B. K. Hofer & P. R. Pintrich (Eds.), *Personal epistemology: The psychology of beliefs about knowledge and knowing*. London: Lawrence Erlbaum Associates.

Schommer, M. (1994). An emerging conceptualization of epistemological beliefs and their role in learning. In R. Garner & P. Alexander (Eds.), *Beliefs about text and instruction with text*. Hillsdale, NJ: Lawrence Erlbaum Associates.

Seale, C. (2000). Using computers to analyse qualitative data. In D. Silverman (Ed.), *Doing qualitative research: A practical handbook*. London: Sage.

Secada, W. G. (1995). Social and critical dimensions for equity in mathematics education. In W. G. Secada, E. Fennema & L. Adajian (Eds.), *New directions for equity in mathematics education*. Cambridge: Cambridge University Press.

Seldon, A., & Seldon, J. (2003). Validations of proofs considered as texts: Can undergraduates tell whether an argument proves a theorem? *Journal for Research in Mathematics Education, 34*, 4–36.

Seymour, E., & Hewitt, N. (1997). *Talking about leaving: Why undergraduates leave the sciences* Boulder, CO: Westview Press.

Sfard, A. (2000). On reform movement and the limits of mathematical discourse. *Mathematical Thinking and Learning, 2*(3), 157–189.

Sfard, A. (2006). Participationist discourse on mathematics learning. In J. Maasz & W. Schloeglmann (Eds.), *New mathematics education research and practice* (pp. 153–170). Rotterdam: Sense.

Sfard, A., & Lavie, I. (2005). Why cannot children see as the same what grown-ups cannot see as different? – Early numerical thinking revisited. *Cognition and Instruction, 23*(2), 237–309.

Sfard, A., & Prusak, A. (2005). Telling identities: In search of an analytic tool for investigating learning as a culturally shaped activity. *Educational Researcher, 34*(4), 14–22.

Shaw, J. (forthcoming). Appetite and anxiety: The mathematics curriculum and "object relations" theory. In L. Black, H. Mendick & Y. Solomon (Eds.), *Mathematical relationships in education: Identities and participation*. New York: Routledge.

She, H.-C. (2000). The interplay of a Biology teacher's beliefs, teaching practices and gender-based student–teacher classroom interactions. *Educational Research and Evaluation, 42*(1), 100–111.

Shield, M., & Galbraith, P. (1998). The analysis of student expository writing in mathematics. *Educational Studies in Mathematics, 36*, 29–52.

Sierpinska, A. (1994). *Understanding in Mathematics*. London: Falmer Press.

Silver, E. A., Smith, M. S., & Nelson, B. S. (1995). The QUASAR project: Equity concerns meet mathematics education reform in the middle school. In W. G. Secada, E. Fennema & L. Adajian (Eds.), *New directions for equity in mathematics education* (pp. 9–56). Cambridge: Cambridge University Press.

Sinclair, J., & Coulthard, M. (1975). *Toward an analysis of discourse: The English used by teachers and pupils*. Oxford: Oxford University Press.

Skinner, D., Valsiner, J., & Holland, D. (2001, September). Discerning the dialogical self: A theoretical and methodological examination of a Nepali adolescent's narrative [Electronic Version]. *Forum: Qualitative Social Research, 2*. Retrieved July 31, 2007 from http://www.qualitative-research.net/fqs-texte/3–01/3–01skinneretal-e.htm

Slavin, R. E. (1990). Achievement effects of ability grouping in secondary schools: A best evidence synthesis. *Review of Educational Research, 60*, 471–490.

Smith, C. (2006). *eNRICH mathematics project evaluation interim report*. Cambridge, UK: Cambridge University.

Snyder, I., & Lankshear, C. (2000). *Teachers and technoliteracy: Managing literacy, technology and learning in schools*. St Leonard's, NSW: Allen & Unwin.

Solomon, Y. (1989). *The practice of mathematics*. London: Routledge.

Solomon, Y., & Black, L. (2008). Talking to learn and learning to talk in the mathematics classroom. In N. Mercer & S. Hodgkinson (Eds.), *Exploring talk in school: Inspired by the work of Douglas Barnes*. London: Sage.

Sowder, L., & Harel, G. (2003). Case studies of mathematics majors' proof understanding, production and appreciation. *Canadian Journal of Science, Mathematics and Technology Education, 3*(2), 251–267.

Spender, D. (1980). *Man-made language*. London: Pandora Press.

Spender, D., & Sarah, E. (1980). *Learning to lose: Sexism and education*. London: Women's Press.

Stanley, J. (1993). Sex and the quiet schoolgirl. In P. Woods & M. Hammersley (Eds.), *Gender and ethnicity in schools: Ethnographic accounts*. Milton Keynes: Open University Press.

Stanworth, M. (1983). *Gender and schooling: A study of sexual divisions in the classroom*. London: Hutchinson.

Star, S. L., & Griesemer, J. R. (1989). Institutional ecology, "translations" and boundary objects: Amateurs and professionals in Berkeley's Museum of Vertebrate Zoology, 1907–39. *Social Studies of Science, 19*, 387–420.

Steele, C. M. (1997). A threat in the air. *American Psychologist, 52*(6), 613–629.

Stein, M. K., Smith, M. S., Henningsen, M. A., & Silver, E. A. (2000). *Implementing standards-based mathematics instruction: A casebook for professional development*. New York: Teacher College Press.

Steinbring, H. (2005). Analyzing mathematical teaching-learning situations—the interplay of communicational and epistemological constraints. *Educational Studies in Mathematics, 59*, 313–324.

Stetsenko, A., & Arievitch, I. M. (2004). The self in cultural-historical activity theory: Reclaiming the unity of social and individual dimensions of human development. *Theory and Psychology, 14*(4), 475–503.

Steward, S., & Nardi, E. (2002). I could be the best mathematician in the world. . .if I actually enjoyed it. *Mathematics Teaching*(180), 4–9.

Stinson, D. W. (2006). African American male adolescents, schooling (and mathematics): Deficiency, rejection, and achievement. *Review of Research in Education, 76*(4), 447–506.

Street, B. (2003). What's "new" in New Literacy Studies? Critical approaches to literacy in theory and practice. *Current Issues in Comparative Education, 5*(2), 77–91.

Strutchens, M. E. (2000). Confronting beliefs and stereotypes that impede the mathematical empowerment of African American students. In M. E. Strutchens, M. L. Johnson & W. F. Tate (Eds.), *Changing the faces of mathematics: Perspectives of African Americans* (pp. 7–14). Reston, VA: National Council of Teachers of Mathematics.

Swain, J., Newmarch, B., Baker, E., Holder, D., & Coben, D. (2004, July 4–11). *Changing adult learners' identities through learning numeracy.* Paper presented at the International Conference on Mathematics Education 10, Copenhagen, Denmark.

Swann, J. (1994). What do we do about gender? In B. Steirer & J. Maybin (Eds.), *Language, literacy and learning in educational practice.* Clevedon: Multilingual Matters.

Swann, J., & Graddol, D. (1988). Gender inequalities in classroom talk. *English in Education, 22,* 48–65.

Tharp, R., & Gallimore, R. (1991). A theory of teaching as assisted performance. In P. Light, S. Sheldon & M. Woodhead (Eds.), *Learning to think.* London: Routledge.

Tizard, B., & Hughes, M. (1984). *Young children learning.* London: Methuen.

Van Zoest, L. R., & Bohl, J. V. (2005). Mathematics teacher identity: A framework for understanding secondary school mathematics teachers' learning through practice. *Teacher Development, 9*(3), 315–345.

van Zoest, L. R., Ziebarth, S. W., & Breyfogle, M. L. (2002). Self-perceived and observed practices of secondary school mathematics teachers. *Teacher Development, 6*(2), 245–268.

Veel, R. (1999). Language, knowledge and authority in school mathematics. In F. Christie (Ed.), *Pedagogy and the shaping of consciousness: Linguistic and social processes* (pp. 185–216). London: Continuum.

Vygotsky, L. S. (1978). *Mind in society.* Cambridge, MA: Harvard University Press.

Vygotsky, L. S. (1987). Thinking and speech (N. Minick, Trans.). In R. W. Rieber & A. S. Carton (Eds.), *The collected works of L. S. Vygotsky. Volume 1:Problems of general psychology* (pp. 39–285). New York: Plenum.

Walkerdine, V. (1988). *The mastery of reason: Cognitive development and the production of rationality.* London: Routledge.

Walkerdine, V. (1997). Difference, cognition and mathematics education. In A. Powell & M. Frankenstein (Eds.), *Ethnomathematics* (pp. 201–214). New York: State University of New York.

Walkerdine, V. (1998). *Counting girls out* (2nd ed.). London: Falmer.

Weber, K. (2001). Student difficulty in constructing proofs: The need for strategic knowledge. *Educational Studies in Mathematics, 48*(1), 101–119.

Wells, G. (1987). *The meaning makers.* London: Hodder & Stoughton.

Wells, G. (1992). The centrality of talk in education. In K. Norman (Ed.), *Thinking Voices: The work of the National Oracy Project.* London: Hodder & Stoughton.

Wells, G. (1999). *Dialogic inquiry: Toward a sociocultural practice and theory of education.* New York: Cambridge University Press.

Wells, G., & Arauz, R. M. (2006). Dialogue in the classroom. *Journal of the Learning Sciences, 15*(3), 379–428.

Wenger, E. (1998). *Communities of practice: Learning, meaning and identity.* Cambridge: Cambridge University Press.

White, D. (2003). Promoting productive mathematical classroom discourse with diverse students. *Journal of Mathematical Behavior, 22,* 37–53.

White, J. (1987). *The language of science: Making and interpreting observations.* London: Assessment of Performance Unit, Department of Education and Science (in collaboration with G. Welford).

Wiliam, D., & Bartholomew, H. (2004). It's not which school but which set you're in that matters: The influence of ability grouping practices on student progress in mathematics. *British Educational Research Journal, 30*(2), 279–293.

Williams, G. (1998). Children entering literate worlds: Perspectives from the study of textual practices. In F. Christie & R. Misson (Eds.), *Literacy and schooling.* London: Routledge.

Williams, S., & Baxter, J. (1996). Dilemmas of discourse-oriented teaching in one middle school mathematics classroom. *The Elementary School Journal, 97,* 21–38.

Wood, D. (1986). Aspects of teaching and learning. In M. Richards & P. Light (Eds.), *Children of social worlds.* Cambridge: Polity Press.

Wood, D., Bruner, J., & Ross, G. (1976). The role of tutoring in problem solving. *Journal of Child Psychology and Psychiatry, 17*, 89–100.

Yackel, E., & Cobb, P. (1996). Sociomathematical norms, argumentation, and autonomy in mathematics. *Journal for Research in Mathematics Education, 27*(4), 458–477.

Zevenbergen, R. (2000). "Cracking the code" of mathematics classrooms: School success as a function of linguistic, social and cultural background. In J. Boaler (Ed.), *Multiple perspectives on mathematics teaching and learning* (pp. 201–223). Westport, CT: Ablex.

Zevenbergen, R. (2001). Language, social class and underachievement in mathematics. In P. Gates (Ed.), *Issues in mathematics teaching*. London: Routledge.

Zevenbergen, R. (2005). The construction of a mathematical habitus: Implications of ability grouping in the middle years. *Journal of Curriculum Studies, 37*(5), 607–619.

Author Index

Subject Index

minority ethnic groups 49, 57, 137–41, 145, 149, 155; African American 48, 140, 141, 144–6, 194–5; African Caribbean 40; Hispanic 139, 141, 145; indigenous Australian 138, 175

motivation 38, 62, 87, 91, 109, 130, 198

narrative: mathematics within 165–76, 214; of self 121–4, 140–1; *see also* genre, identity, positioning

National Numeracy Strategy 11, 203

neo-Vygotskian 15, 18, 81, 180

novice 23, 85, 105, 126

NRICH 190

pedagogy: discourse 14, 30–5, 50–4; inclusive 38, 130, 183, 197–9; practice 7, 8, 177, 194; reform 38, 42, 48, 58, 60, 140, 146, 160, 189, 193, 197; traditional 22, 91, 119, 131, 134, 137, 146, 150

peer: learning *see* collaborative learning, classroom interaction

performance 22, 37, 40–2, 52, 57–9, 74–9, 100–1, 105, 109–12, 134, 146, 153, 160, 202, 203

positioning: by others, 7, 58, 121, 122; of self 18, 20, 38, 48, 50, 54–60, 70–2, 76, 79, 90, 93, 107, 117–18, 140, 150, 155, 191

power 16, 25–6, 32, 48, 54, 107, 128–30, 137, 161, 165, 168, 177–84

primary school 4–7, 13–14, 16, 27, 29, 32–49, 140, 155–9, 201–6; *see also* elementary school

problem-solving 17, 59, 73, 112, 169, 176, 189–90, 194

Project Impact 191

proof 22, 96–7, 110–19, 160–1, 175; *see also* argument

pure mathematics 96, 101–2, 108–9, 151

questions *see* classroom interaction

QUASAR 191

realistic *see* formal versus informal contexts

real world problems *see* formal versus informal contexts

reform, mathematics education 42, 58, 60, 140, 160, 189, 193, 197; *see also* pedagogy

register, mathematics 9, 11, 50, 138–9, 164–6, 177, 181, 192–5, 213–14

relational classrooms 130, 132, 197–9

resistance 7, 25, 48, 51–4, 80–2, 122, 132–3, 141–2, 160, 197, 199

rule-following in mathematics 18, 73, 78, 94–5, 105, 116, 118, 143, 154, 176, 180, 195, 211

scaffolding 15, 45, 158, 166

secondary school 16, 27–8, 51, 56–60, 61–83, 148, 152, 159, 180, 201–5, 207

setting 56–9, 63–4, 67–71, 78–81, 152–5, 159; *see also* ability

social class 12, 31–5, 53, 56–7, 138–44, 149, 159, 166, 177–8, 191

speed 7, 22, 41–3, 62–4, 67, 70–2, 79, 87–90, 100–2, 110, 126, 154, 159, 177

statistics, studying 108–9

stereotypes 25, 54–6, 92, 108, 124, 131, 141, 149; *see also* geek, genius, positioning, identity

Student Experiences of Undergraduate Mathematics project 86–93, 105, 152

student support center *see* collaborative learning

success 12–15, 38, 55–6, 62, 68, 70, 72, 86, 96, 99, 105–6, 110, 118–21, 125–6, 140–1, 198–200

textbooks 78, 138, 142, 176; *see also* language

third space 197–8

tier system, in GCSE 71, 140–3, 263

trust 179, 182–5, 195, 199

under-achievement 12, 40, 137

undergraduates: community 89–90, 98–107, 118; student-lecturer relationships 114–17, 125–30

Milton Keynes UK
Ingram Content Group UK Ltd.
UKHW022106141024
449569UK00031B/1803